C000108728

MATHEMATICAL MODELLING IN EDUCATION AND CULTURE: ICTMA 10

When we mean to build
We must first survey the plot, then draw the model;
And when we see the figure of the house,
Then must we rate the cost of the erection.
Which if we find outweighs ability,
What do we then but draw anew the model
In fewer offices, or, at least, desist
To build at all?

Shakespeare on modelling: *Henry IV, Pt 2*

Mathematics possesses not only the truth, but supreme beauty - a beauty cold and austere like that of stern perfection, such as only great art can show.

Bertrand Russell (1872-1970) in *The Principles of Mathematics*

ABOUT THE EDITORS

Qi-Xiao Ye, Professor of Mathematics, completed his graduate studies at Peking University, Beijing, China in 1963. His research field is partial differential equations, especially reaction-diffusion equations. He was heavily involved in both the modelling and designing of a ship lock in Gezhou Dam on the Yangtze River and also in the design and implementation of the associated software. He has taught many courses on mathematical modelling, is one of the chief organisers of the China Undergraduate Mathematical Contest in Modelling (CUMCM) and has a strong interest in the reform of mathematics education at university and high school level.

Werner Blum gained his PhD in Pure Mathematics and is currently Professor of Mathematics and Mathematical Education, University of Kassel, Germany. His research interests include mathematical modelling and applications, mathematical literacy and quality of instruction, strategy in mathematics teaching and mathematics in vocational education. He has been Co-Chairman of the Kassel-Exeter project and President of the GDM. He is currently Chairman of the ICMI Study on Applications and Modelling in Mathematics. He is a member of the PISA Expert Group and chairs a model project for quality development.

Ken Houston, Professor of Mathematical Studies in the School of Computing and Mathematics, University of Ulster, gained his BSc Honours and PhD degrees from Queen's University, Belfast. He has long been involved with ICTMA and is currently President. He has a strong interest in the teaching, learning and assessment of mathematical modelling and has served on several national and international committees relating to mathematics and mathematics education.

Qi-Yuan Jiang, Professor of Mathematics, graduated from Tsinghua University, Beijing, China in 1962. His research interests include mathematical modelling and applications and operations research. He has taught courses on mathematical modelling since the early 1980s, and at present he teaches a course on experiments in mathematics. He is one of the chief organisers of the China Undergraduate Mathematical Contest in Modelling (CUMCM) and has a strong interest in the reform of mathematics education.

MATHEMATICAL MODELLING IN EDUCATION AND CULTURE: ICTMA 10

Editors:

Qi-Xiao Ye
Beijing Institute of Technology
China

Werner Blum
University of Kassel
Germany

Ken Houston
University of Ulster
Northern Ireland

Qi-Yuan Jiang
Tsinghua University
China

Horwood Publishing
Chichester

First published in 2003 by
HORWOOD PUBLISHING LIMITED
Coll House, Westergate, Chichester, West Sussex, PO20 6QL England

COPYRIGHT NOTICE
All Rights Reserved. No part of this publication may be reproduced, stored in a retrieval system, or transmitted, in any form or by any means, electronic, mechanical, photocopying, recording, or otherwise, without the permission of Horwood Publishing Limited.

© Horwood Publishing Limited, 2003

British Library Cataloguing in Publication Data
A catalogue record of this book is available from the British Library

ISBN 1-904275-05-2

Printed in Great Britain by Antony Rowe Limited

Table of Contents

Preface   x

Section A - Research in Teaching, Learning and Assessment

1   Context in application and modelling – an empirical approach   3
Andreas Busse[1] and Gabriele Kaiser[2]
[1]In-service Teacher Training Institute, Hamburg, Germany
[2]University of Hamburg, Germany

2   Mathematical modelling as pedagogy: Impact of an immersion programme   16
Trudy Dunne[1] and Peter Galbraith[2]
[1]Downlands College, Toowoomba, Queensland, Australia
[2]The University of Queensland, Australia

3   Using ideas from physics in teaching mathematical proofs   31
Gila Hanna[1] and Hans Niels Jahnke[2]
[1]Ontario Institute for Studies in Education, The University of Toronto
[2]Universität Essen, Germany

4   Deconstructing mathematical modelling: Approaches to   41
     problem solving
     Christopher Haines[1], Rosalind Crouch[2] and Andrew
     Fitzharris[2]
     [1]City University, London, United Kingdom
     [2] University of Hertfordshire, Hatfield, United Kingdom

5   Investigating students' modelling skills   54
     Ken Houston and Neville Neill
     University of Ulster, Northern Ireland

6   "How to model mathematically" table and its   67
     applications
     Wang Geng
     Nanjing University of Economics, China

Section B - Mathematical Modelling Competitions

7   New applications of the mathematics A-lympiad   81
     Dédé de Haan
     Freudenthal Institute, Utrecht University, The
     Netherlands

8   Mathematics contest in modelling: Problems from   93
     practice
     Shang Shouting, Zheng Tong and Shang Wei
     Harbin Institute of Technology, China

Section C - Using Technology in the Teaching of Modelling

9   Modelling and spreadsheet calculation   101
     Mike Keune[1] and Herbert Henning[2]
     [1]Ecumenical Cathedral School of Magdeburg, Germany
     [2]Otto-von-Guericke University of Magdeburg, Germany

10 Technology-enriched classrooms: Some implications for 111
teaching applications and modelling
Peter Galbraith, Peter Renshaw, Merrilyn Goos and
Vince Geiger
University Of Queensland, Australia

11 Choosing and using technology for secondary 126
mathematical modelling tasks: Choosing the right peg
for the right hole
Vince Geiger, Peter Galbraith, Peter Renshaw and
Merrilyn Goos
University Of Queensland, Australia

Section D - Models for Use in Teaching

12 Groups, symmetry and symmetry breaking 143
Albert Fässler
Hochschule für Technik und Architektur HTA,
Biel/Bienne, Switzerland

13 The rainbow: from myth to model 153
Hans-Wolfgang Henn
University of Dortmund, Germany

14 Teaching inverse problems in undergraduate level 165
mathematics, modelling and applied mathematics
courses
Fengshan Liu
Delaware State University, Dover, U.S.A

15 Bezier curves and surfaces in the classroom 173
Baoswan Dzung Wong
Aargauische Kantonsschule Wettingen, Switzerland

Section E - Teacher Education

16   A mathematical modelling course for pre-service   183
secondary school mathematics teachers
*Zhonghong Jiang, Edwin McClintock and George
O'Brien*
Florida International University, USA

17   Mathematical modelling in teacher education   197
Mikael Holmquist and Thomas Lingefjärd
Göteborg University, Sweden

18   Two modelling topics in teacher education and training   209
Adolf Riede
University of Heidelberg, Germany

Section F - Innovative Modelling Courses

19   The knowledge and implementation for the course of   225
mathematical experiment
Zhao Jing, Jiang Jihong, Dan Qi and Fu Shilu
Logistics Engineering College, Chongqing, China

20   Teaching patterns of mathematical application and   233
modelling in high school
Tang Anhua, Sui Lili and Wang Xiaodan
Beijing No.15 High School, China

21   Mathematical experiment course: Teaching mode and   249
its practice
Qiongsun Liu, Shanqiang Ren, Li Fu and Qu Gong
Chongqing University, China

22   The mathematical modelling – orientated teaching   257
method of elicitation
Ruiping Hu and Shuxia Zhang
Zhenjiang Watercraft College of CPLA, Jiangsu, China

23   Teaching and assessment of mathematical modelling in   267
        community colleges
        Lu Xiuyan, Mo Jingjing and Lu Keqiang
        Feng Tai Workers and Staff University, Beijing, China

24   The role of mathematical experiment in mathematics   284
        teaching
        Jinyuan-Li
        Beijing No.4 High School, China

25   Theory and practice in teaching of mathematical   293
        modelling at high school level
        Qiu Jinjia
        The High School Affiliated To Renmin University Of
        China

Section G - Panel Discussion

26   Report of panel discussion   307
        Chair:
        Peter Galbraith, University of Queensland, Australia
        Panel:
        Ken Houston, University of Ulster, Northern Ireland
        Tomas Jensen, Roskilde University, Denmark
        Gabriele Kaiser, University of Hamburg, Germany
        Yongji Tan, Fudan University, China

Preface

The 10th International Conference on the Teaching of Mathematical Modelling and Applications (ICTMA 10) was held in Beijing, in July/August 2001. This was the first time that the conference had been held in an eastern country, and it was a timely move. As well as the political and cultural changes evident in China over the last number of years, there are also the beginnings of important changes to the mathematics curricula in both schools and universities. The influence of two decades of the mathematical modelling movement in the west was beginning to be felt, and both teachers and academics were eager to meet their western counterparts and to share with them their excitement at these developments.

This volume contains a collection of 25 of the contributions to ICTMA 10, and a high percentage of these are by Chinese authors who describe their developing work. The rest of the contributions are by authors resident in "the west", including Australia, and these papers describe ongoing developments in curriculum design and teaching methods. However they also concentrate on research into the learning of modelling, and this research is giving the community greater insights into the ways in which students develop as modellers. These insights will inform pedagogy, making it more effective and efficient.

There is an innovation in this book. The plenary panel discussion session was recorded, transcribed and edited into a report by the chair of the session. This is included as Chapter 26 and it gives some insights into the thinking of the "experts" who were foolhardy enough to sit on the panel!

Each chapter is assigned to one of seven themed sections, but the reader should be aware that the allocation of chapters to themed sections is an inexact science (apart from assigning chapter 26 to section G, that is).

Accordingly they should look carefully at all chapters to ensure that they do not miss anything of interest.

Section A has six chapters dealing with various research programmes. The two-chapter section B describes the role that competitions for students can play in developing their modelling skills. Technology continues to develop and to play a big part in modelling and the three chapters in section C describe both innovation and research in the use of technology in teaching modelling. The four chapters in section D provide exemplars to enrich teaching while section E has three chapters of particular interest to teacher educators. However the most culturally enriching section is section F, which has seven chapters describing developments in China.

The book as a whole complements the series of books emanating from the biennial ICTMA conferences (see next page for details) and it aims to provide an additional source of information and ideas for those engaged in the teaching, learning and assessment of mathematical modelling in secondary and tertiary institutions. It will be of particular interest to any who want to read about recent developments in China.

Author's affiliations and e-mail addresses are given in each chapter to facilitate communication. We are grateful to all authors for their contributions, their care in presenting their material and their patience while waiting for the editors to complete the book.

We thank the International Executive Committee of the ICTMA and the Local Organising Committee in Beijing for the roles they played in making the conference happen. Finally we are grateful for the continued support and encouragement of our publisher Ellis Horwood (Horwood Publishing) who has been a consistent supporter of the ICTMA series over many years.

We hope that the reader will enjoy this book and will find it useful for enhancing their teaching, for stimulating their research or for informing them of world-wide developments.

The ICTMA website is http://www.infj.ulst.ac.uk/ictma/

Qi-Xiao Ye   Werner Blum   Ken Houston   Qi-Yuan Jiang

BOOKS EMANATING FROM ICTMA CONFERENCES

Berry JS, Burghes DN, Huntley ID, James DJG and Moscardini AO, 1984, *Teaching and Applying Mathematical Modelling*, Ellis Horwood, Chichester.

Berry JS, Burghes DN, Huntley ID, James DJG and Moscardini AO, 1986, *Mathematical Modelling Methodology, Models and Micros*, Ellis Horwood, Chichester.

Berry JS, Burghes DN, Huntley ID, James DJG and Moscardini AO, 1987, *Mathematical Modelling Courses*, Ellis Horwood, Chichester.

Blum W, Berry JS, Biehler R, Huntley ID, Kaiser-Messmer G and Profke L, 1989, *Applications and Modelling in Learning and Teaching Mathematics*, Ellis Horwood, Chichester.

Niss M, Blum W and Huntley ID, 1991, *Teaching of Mathematical Modelling and Applications*, Ellis Horwood, Chichester.

De Lange J, Keitel C, Huntley ID and Niss M, 1993, *Innovation in Maths Education by Modelling and Applications*, Ellis Horwood, Chichester.

Sloyer C, Blum W and Huntley ID, 1995, *Advances and Perspectives in the Teaching of Mathematical Modelling and Applications*, Water Street Mathematics, Yorklyn, Delaware.

Houston SK, Blum W, Huntley ID and Neill NT, 1997, *Teaching and Learning Mathematical Modelling*, Albion Publishing Ltd (now Horwood Publishing Ltd.), Chichester.

Galbraith P, Blum W, Booker G and Huntley ID, 1998, *Mathematical Modelling, Teaching and Assessment in a Technology-Rich World*, Horwood Publishing Ltd., Chichester.

Matos JF, Blum W, Houston SK and Carreira SP, 2001, *Modelling and Mathematics Education ICTMA 9: Applications in Science and Technology,* Horwood Publishing, Chichester.

Ye Q, Blum W, Houston SK and Jiang Q, 2003, *Mathematical Modelling in Education and Culture: ICTMA 10*, Horwood Publishing, Chichester [this volume].

Lamon S, Parker W and Houston K, 2003, *Mathematical Modelling: A Way of Life – ICTMA 11*, Horwood Publishing, Chichester.

Section A

Research in Teaching, Learning and Assessment

1

Context in Application and Modelling – an Empirical Approach

Andreas Busse,
In-service Teacher Training Institute, Hamburg, Germany,
busse@ifl-hamburg.de

Gabriele Kaiser,
University of Hamburg, Germany,
gkaiser@erzwiss.uni-hamburg.de

Abstract

In a qualitative-orientated empirical study the effect of context on the students' approach towards real world problems is investigated. First results show that contexts given in a task can internally be reconstructed in very different ways. It is also shown that both, the student's affective proximity to the context and the student's real world knowledge can have fostering but also hindering effects on his or her performance.

1. Introduction

The theme of the paper is context, a rather nebulous concept, used by many authors in different meanings and ways. Sometimes even different concepts of context can be found in the same study. Also several names are given for this concept such as cover story, situation, situational context, setting, social context, etc.

On the other hand there is consensus that context - however it is meant - is a central topic in the discussion about modelling and applications in mathematics education. Since its very beginning the debate on the teaching of mathematical modelling and applications claims the usage of various contexts or contextualized problems. We associate many goals with the usage of contextualized problems, among others the fostering of motivation for the learning of mathematics or as a way to visualize mathematical concepts and methods (Kaiser, 1995). Until now we do not have comprehensive empirical evidence how far our hopes concerning the usage of contextualized problems are fulfilled.

The empirical study we are going to present tries to give a few answers to this problem, although we are far away from having final answers. First we will clarify the meaning of context. Afterwards some results from previous empirical research studies are presented. In the third part the aim of our own study and the methodical design will be described. In the fourth part first observations are detailed, finally in the fifth part first results of the influence of the so-called *subjective figurative context* will be described.

2. Research results from previous studies

Clarke and Helme (1996) propose to distinguish between *figurative context* and *interactive context*. *Figurative contex*t comprises the real scenario the task is embedded in. *Interactive context* describes the conditions the task is encountered in by a student. Applying a constructivist perspective they point out that both kinds of context are individually constructed.

Following this idea in our study we distinguish between

objective figurative context: description of the real scenario given in the task

subjective figurative context: individual   interpretation   of   the *objective figurative context*

Although a similar distinction is theoretically also possible for the *interactive context,* we avoid this distinction for reasons of clarity since the focus of this paper is figurative context.

It is presumably the *objective figurative context* that is often implicitly meant by researchers when referring to "context".

Objective figurative contexts are generally chosen so that they are presumably near to students. Generally this is linked to the expectation of a fostering effect for the problem-solving process. However, which effect the objective figurative context offered in the task does have on a student depends on the way he or she constructs his or her own *subjective figurative context* (see as an example the story about the dog Nadja in section 3). The effects of an objective figurative context have not been investigated comprehensively yet, the studies carried out so far do not provide a uniform picture.

In the research results of almost all of the empirical studies on context the term *figurative context* is not used by researchers, rather many different expressions are used - among others the word context. For the sake of clarity we use the term *figurative context* in the meaning of *objective* figurative context in this paper unless stated otherwise.

The culture of the mathematics classroom seems to be a factor. Boaler (1993) found in her comprehensive study that in an open, discussion-centred mathematics-classroom the figurative context had a smaller effect than in traditional classrooms where figurative contexts seem to influence the students' performance. But no general pattern could be found which figurative contexts have a fostering and which have a hindering effect.

According to Stillman (1998) the degree of involvement with the figurative context is influenced by the task type. So-called "wrappper problems" provoke less, modelling-tasks more involvement with the figurative context. Generally she found that medium to high involvement with the figurative context leads to better performance. But there are exceptions from this tendency, especially concerning students who perform generally well in mathematics.

When investigating the problems minority students in the Netherlands have with contextualized tasks Carvalho de Figueirdo (1999) found out that language and cultural background have an effect on the way figurative contexts are understood. It was also found that illustrations can mislead students in their understanding of the figurative context of a task. Furthermore it was pointed out that an unfamiliar figurative context tempt students to de-contextualise the task quickly.

In some research studies not presented here *familiarity* with the figurative context is mentioned as a fostering effect on the students' performance. But this is not generally confirmed in other studies.

To sum it up: In all the studies mentioned it remains an open question how the figurative context offered in the tasks (by words or illustrations) is represented internally and which effects it shows in the course of the solution process.

These questions were the central questions of our own qualitative-orientated study.

3. Aim of the study and methodical design

The aim of the study is the generation of hypotheses about the effects the figurative context can have on secondary school students. We are aiming at a development of a categorization of the effects observed.

This study uses the case study approach, therefore generalizations cannot be made easily and have to reflect the special characteristics of the sample and the observation design.

The development of the methodical design of the study was rather difficult, because we needed methods which allowed us to reconstruct the internal representations of the context by an individual student and which gave us insight into the emotional reactions of the student. The problem was that no direct observation or measurement was possible.

We carried out some pre-studies. Among other approaches a group of students was asked to write free stories about the task immediately after they had worked on it. This was done with 12- and 19-year-old students.

Let us give an example of the story a twelve year girl wrote after working on the following task:

A dog's first year is equivalent to 15 human years. The second year of a dog is equivalent to 6 human years. Each of the following years of a dog is equivalent to 5 human years.

(After this introductory text some calculation-tasks were given.)

After solving the task correctly the girl wrote:

I have a dog named Nadja. She was born in 1986 and her age is 13 dogyears, which is approximately 76 human years. Anyway, when I was approximately 2 years old I was sitting on the baby's changing unit. I could not feel my legs from the first day on. But somehow, nobody knows how, I tried to crawl. Nadja heard that and came running into the living room. I fell down. Fortunately Nadja was there and I fell on Nadja's back and then onto the floor. Up to the age of 3 I could not feel my legs! All that is true!!!

Independent from the truth of the story this is an impressive and touching example of what can be on a student's mind when confronted by a task. This observation, among others, confirmed our conjecture that figurative contexts can have very individual, unpredictable effects. This reinforces the necessity of the concept of the *subjective figurative context* we mentioned in the beginning.

As a result of the pre-studies the following three step design appeared to be the best: Four pairs of students (16 to 17 years old, in their 1st year of upper secondary school, two pairs of boys, two pairs of girls) from four different schools were given realistic tasks. The students were videotaped while working on the tasks. After the work on the task the students individually watched the video-recording. They were requested to stop the tape when during the solution-process something concerning the figurative context crossed their mind. They were asked to utter these thoughts. The statements were audiotaped. The interviewer could as well stop the tape if the student did not do so while the interviewer considered certain parts of the tape as potentially useful for statements. This method is known as method of stimulated recall (e.g. Weidle and Wagner, 1994). In an immediately following interview the student was asked probing questions about the statements made during the stimulated-recall-phase.

The purpose of the design was to catch thoughts, associations, etc. concerning the figurative context in that moment in which they crossed the student's mind ("in statu nascendi") without disturbing the solution process. The statements made during the stimulated-recall-phase were usually not interrupted by the interviewer to avoid disturbances. The place and time for further questions was the interview.

The data sampling was finished early 2001, not all of the 24 data-sets have been transcribed yet. We are currently in the process of a first systematic approach towards the data. So what we can present are first results far from being final. Especially no generalizations of or remarks on the quantitative spread of the observed phenomena can be made.

The tasks were given on three different days. The tasks were matched to different objective figurative contexts. The tasks differed also in the degree of mathematical modelling required. The mathematical theory required for all the tasks was taken from former years of the students' maths-education. There are two reasons for that:

➢ It was not clear which topics had already been covered at the different schools.
➢ The mathematics should remain in the background for the benefit of the figurative context. In other words: The maths was supposed to be well known.

The first task which was given to the students is shown in Table 1. The text is translated from German. Before we refer to the students' results let us comment briefly on the task:

➢   The data are authentic, they were taken from publications of an oil-company. The students had not been informed about the authenticity of the data before they were working on the tasks.

➢   One might be tempted to solve the task by applying a mere extrapolation. But in order to determine the consumption in a certain number of years it is necessary to cumulate the consumption figures. This is the main mathematical difficulty of this task.

➢   The global consumption seems to be almost linear. Indeed if the results of a linear approach are compared with those of an exponential one only little difference is found. Using a linear model one gets a range of 32 years, the exponential approach leads to a range of 31 years. (It might be interesting to know that these figures meet well with predictions made by geo-scientists.)

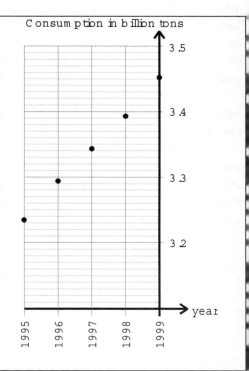

By the end of the year 1999 the global oil reserves were estimated to be approximately 138.041 billion tons.

In recent years the global oil consumption has been increasing steadily.

The consumption figures from the past can be taken from the diagram opposite.

a)   Determine approximately the amount of global oil consumption that can be expected for the year 2000.

b)   When approximately will the reserves be used up?

Table 1

4. Reactions of the students on the task

The task was almost acceptably solved by one pair by using a table. Two pairs used a promising approach but failed in applying the mathematical techniques properly. One pair did not really understand the task.

We do not want to present the answers rather than giving an outlined reconstruction of the statements made about the figurative context. We restricted the number of students in this paper for the sake of clarity. Future publication will show more detailed results.

In the transcripts we present many filling words were omitted for reasons of better understanding, which is shown by dots in brackets. Underlined words were spoken with emphasis. A question mark indicates that the spoken word could not be heard properly.

Christine almost completely ignored the figurative context of the task. Only in the beginning she mentioned a task that appears to be similar for her. This task was about cell growth. She tried by adapting known and looked-up formulae to the task to come to a result. She accepted her result of a range of approximately 361 years without any comment. (Remember: a correct answer would be approximately 32 years.)

During the stimulated recall phase she made no statements about the figurative context (in contradiction to the instructions given) but explained her way of working on the task focussing on technique.

In the interview Christine was asked questions about the figurative context:

Interviewer: Have you ever dealt with this <u>topic</u> in another school subject? Or in another way through newspapers or-
Christine: Hm- for which topic?
Interviewer: What this is about here.
Christine: Mineral oil? Oh, you mean mineral oil?
Interviewer: If you think <u>mineral oil</u> is the topic, so that's what I mean.
Christine: Ehem. No. [laughing]

Later in the interview she said:

Christine: Mineral oil always sounds so uninterestingly.

So for Christine the figurative context hardly seemed to exist when working on the task. Even later in the interview she seemed to think for the first time that the task even <u>has</u> a figurative context. Then (and not during the solution process) she gave contextual comments.

Ingo associated more than Christine. His first thought was the driving licence test his sister had passed on the same day. He reflected his personal situation in the future when oil would be used up. He had an idea about the value of the range until oil will be used up and said that his knowledge was based on media-publications. He talked about alternative technologies which could extend the range of the oil-reserves and also mentioned a long discussion with his chemistry-teacher about the subject. The oil crises and the gulf war was mentioned by him and he said in the interview that he was concerned about the increasing oil-consumption:

> *Ingo*: At any <u>rate</u> I also would worry about, and I think as it is mentioned in the media, it is interesting for <u>every</u> student of my age how it will be in future. (...) And (...) well, I personally <u>worry</u> about that; well, I have already thought about it quite often. That's really in my mind. The point where it will be finished.

He had some problems in applying mathematical techniques properly, so he failed very close to a successful completion of the task.

Karla associated in an emotional way the problem of destruction of natural environments. Her chemistry lessons provided her with knowledge about how oil comes into being, the limitations of natural resources and other ecological problems. She liked chemistry lessons and her chemistry teacher, which she described in the interview as follows:

> *Karla*: Yes that was a topic I <u>am</u> interested in for its ecological aspects. And also chemistry is a school subject I like; I get on well with my chemistry teacher. (...) and therefore, it has been an important topic to me, also in <u>that</u> lesson. And I liked it, and we watched such a film about when energy resources will be exhausted and (...) that actually did interest me and <u>concerned</u> me I would say.

Karla reflected the assumption of continuously increasing oil consumption and was rather considering a decreasing oil consumption. On the background of the emotional importance the problem of the destruction of natural environment had for her, one is almost tempted to interpret this as a wish. She said in the stimulated recall:

> *Karla*: (...) yes, there one notices that the earth is already quite wrecked, and (low voice) in this respect it actually interested me and (louder) then

I thought about it again in that situation. (...) if really every year zero point zero seven (...) billion tons <u>more</u> are consumed or if it is perhaps only for a while or (...) if one really may <u>say</u> it that way, that it always increases or may be it was (only?) (...) for three or five years like that (...) and that then it will decrease again or (...) what might happen somehow to solve the ecological problem (...).

Karla solved the task by using tables satisfyingly with only one little mistake.

Let us summarize: It becomes obvious that different students interpreted the objective figurative context offered in the task in very different ways. The differences appeared in the associated aspects of reality and in emotional attitudes.

It must be mentioned that the students' statements of course do not cover completely their internal representations of the figurative context offered in the task. One has to expect that many thoughts and associations remain unmentioned due to either missing consciousness of the thoughts or to a missing will to tell them.

In spite of the just mentioned limitations the differences in the perception and interpretation of the objective figurative context offered in the tasks are obvious. These differences are based on individual knowledge, experience and attitude. Consequently reflections on the effects of figurative contexts must consider its individual nature.

5. Influence of the subjective figurative context

In the following we want to present first results of possible contextual influences gained by the reconstruction of the *subjective figurative contexts*.

5.1. *The subjective figurative context can cause a distraction from the task.*

Although generally contextualized tasks are seen as helpful, distracting effects could also be observed. For example Ingo mentioned in the interview:

Ingo: (...) somehow has reminded me of statements and such things I read about. (And?) then (...) at first I had to shake off this thought, so that I could concentrate (...) again (on the?) task because (...) for a moment I was tempted to think too <u>much</u> about it.

Interviewer: It would have disturbed you?

Ingo: (answering quickly) Yes.

Interviewer: Because it-

Ingo: Because (...) it does not help us to solve the problem and because then it is difficult to concentrate again on the problem. Because then one has another thought at the same time, and of course one associates personal experiences or what one has read. If one thinks too <u>much</u> about it, it is more difficult, then one must come back (...) to the problem.

This phenomenon of distraction can have different aspects. So an emotional and a more matter-of-fact-related distraction could be found. Let us present an example for each of the two aspects mentioned.

5.1.1. Emotional involvement in the figurative context can superpose or even disturb the work on the task

This is remarkable because usually it is expected that familiarity with or interest in the figurative context of the task has a helpful effect. The opposite seems to be possible as well.

The above mentioned student Karla who showed a very emotional involvement said in the interview that during the work on the task she always had to think about the films about environmental destruction shown in the chemistry lessons. Asked if she felt disturbed by that she answered:

Karla: Yes, may be, it happens often to me that somehow I totally have to think of different things while solving problems, also during tests. (...) And then I say to myself <u>concentrate</u> and so and then it <u>works</u>, but in that moment one thinks about these <u>films</u> and so and then one does not think about the <u>task</u> and on how to solve it. That is a <u>different</u> problem.

This is interesting because there exist didactical approaches to support especially girls by using figurative contexts which are assumed to be close to girls. Further-going analyses have to be done in this field.

5.1.2. A rich subjective figurative context can be a disturbing factor when working on the task

In a certain sense a rich subjective figurative context produces a world too large for the more or less narrow task. The subjective figurative context contains more information than the text of the task itself and so the student has to choose what is important and what is not.

During the problem-solving process and in the stimulated-recall-phase the student Arthur mentioned a large variety of real-world-factors

that could influence the global oil consumption. Asked about that in the interview:

> *Interviewer*: Which influence did these thoughts have (…) on you while working on problems? Can you tell it?
>
> *Arthur*: (...) (low voice) while solving the problem they only made it more difficult because it was (?) difficult to <u>estimate</u>. I (somehow?) assumed a difference of ten years. That is <u>mathematically</u> hardly solvable.
>
> *Interviewer*: Well, I understand that you were thinking so many things how complex it is-
>
> *Arthur*: Yes.
>
> *Interviewer*. And that has disturbed you?
>
> Arthur: (answering quickly) Yes.
>
> *Interviewer*: Hm. Okay.
>
> *Arthur*: Well, I mean it has made me <u>thoughtful</u> because if (...) one <u>imagines</u> how it might be, the problem cannot be solved. If one is given (a formula?) how long it might last. I think after being given the formula a task could be solved <u>easier</u>. But like this it was (low voice) too (...) big field to (?) simply.

Here a distraction from the task caused by too much content information and not by emotional involvement as described above takes place.

Again this seems to be paradox: A task which is meaningful to the student becomes more difficult due to the host of associations.

5.2. Frequently the subjective figurative context is used to control whether a result is plausible. In conjunction with a misconception of the modelling process this can cause an error

In the case we have in mind the student Josef (Arthur's partner) predicted that the global oil consumption for the year 2000 would be *lower* than the consumption for 1999. They used the mean of the consumption-figures for the years 1995 to 1999 as a prediction for 2000. After feeling that this might be incorrect the student finally accepted this result giving the following reason: The result can be correct because the lately introduced German ecological tax leads to decreasing oil consumption.

Later asked in the interview how he would argue if the figurative context was ore deposits the student said in <u>this</u> case he would not accept the result because he did not know anything about ore deposits.

How can this be understood? The following explanation is possible: Josef had - probably not consciously - build and evaluated a mathematical model. After the evaluation (and not, as it would be correct, while building the model) he added an additional figurative-context-assumption, the introduction of the ecological tax. So he justified a result, which appeared not to be correct at the first sight.

He did not, as it would be correct, start another cycle of the modelling process including the new assumption of the introduction of ecological taxes. Here a kind of modelling- error took place, to which the student was tempted by the figurative context of the task. It becomes clear, how important it is to talk about mathematical modelling in the mathematics-classroom also on a meta-level to avoid mistakes like this.

5.3. The subjective figurative context can cause motivation.

Although this seems to be common-sense it should be mentioned here after all the negative effects we mentioned earlier in this paper. Motivation is an effect which is sometimes taken as the only one. It remains an open question at the moment how strong the motivational effect is, if it carries through the work on the task or if it is just a mind-opener in the sense that it serves as a catcher.

The above mentioned student Ingo stated in the interview:

> *Ingo*: [..] had the (...) problem <u>without</u> theme, just purely mathematical, well, <u>naked</u>, then for instance I would have been sitting down and thinking no, don't feel like doing it. Well, in some way I would not have gone <u>into</u> the problem, very probably. (...) That this topic has helped me <u>very</u> much to find a way for getting into the mathematics. Otherwise I would never have come to this point where I worked with enthusiasm on it.

6. Final remarks

The hopes concerning the inclusion of contextual problems such as the fostering of motivation and understanding claimed in the theoretical debate on applications and modelling have to been seen in a more differentiated light. We have to consider that the figurative context might have a variety of influences, which might be fostering or hindering. We even have to consider that the same figurative context might have different effects on a student dependent on his or her foregoing experiences or the situation the student is in. We have to be more careful with psychological arguments concerning the inclusion of modelling and applications in mathematics education. We should put more emphasis on normative reasons for the inclusion of real world problems in mathematics education. Furthermore consequences for teacher education and in-service-training have to be taken: We need to make student-teachers and practising teachers more aware of the students' various individual reconstructions of the figurative context offered in a task and the possible influences a figurative context can have on the students.

References

Boaler J (1993) 'Encouraging the Transfer of "School" Mathematics to the "Real World" Through the Integration of Process and Content, Context and Culture' *Educational Studies in Mathematics* 25, 341-373.

Carvalho de Figueirdo N J (1999) 'Ethnic minority students solving contextual problems', Utrecht: Freudenthal Institute.

Clarke D and Helme S (1996) 'Context as Construction' *Proceedings CIEAM* 47, 379-389.

Kaiser G (1995) 'Realitätsbezüge im Mathematikunterricht - Ein Überblick über die aktuelle und historische Diskussion' in Graumann G et al (Eds.) *Materialien für einen realitätsbezogenen Mathematikunterricht* Vol. 2, Bad Salzdetfurth: Franzbecker, 66-84.

Stillman G (1998) 'Engagement with task context of application tasks: student performance and teacher beliefs' *Nordic Studies in Mathematics Education* 6, 3/4, 51-70.

Weidle R and Wagner A (1994) 'Die Methode des Lauten Denkens' in Huber L et al (Eds) *Verbale Daten*, 2nd ed, Weinheim,: Beltz, Psychologie Verlags Union, 81-103.

2

Mathematical Modelling as Pedagogy – Impact of an Immersion Program

Trudy Dunne
Downlands College, Toowoomba, Queensland 4350, Australia
dunnet@downlands.qld.edu.au

Peter Galbraith
The University of Queensland, Queensland 4072, Australia
p.galbraith@mailbox.uq.edu.au

Abstract

While mathematical modelling has been incorporated in a variety of ways in school curricula it appears that efforts in which it has been chosen as the defining approach are rare. This is certainly true of courses in the junior secondary school. The purpose of the study reported here was to investigate the impact of a systematic teaching program, in which the development of concepts and skills, as well as the ability to apply mathematics, was pursued by means of mathematical modelling focused around a sequence of carefully selected problems. This paper reports on aspects of program design, teaching methodology, response of students to the program, and performance outcomes.

1. Background

Arguments for the inclusion of mathematical applications in curricula have been appearing for some time, and these may be regarded as 'official' recognition of arguments advancing the ability to apply mathematics to real-life (or life-like) situations as a central goal for mathematics learning (Blum and Niss, 1991). A problem with some reported initiatives has been the brief exposure afforded the modelling emphasis, which does not provide a context for assessing the impact of a major shift in pedagogical style and purpose. This paper reports on an extended initiative that was designed as an 'immersion' program. That is the substance of an entire curriculum was embedded in a modelling approach. This program was implemented at year 8 level, and involved male students of 12 years of age in their first year of secondary education. The context was a single-sex college in a provincial city in Queensland, Australia.

2. Description of the Study

The purpose of the study was to investigate the impact of a systematic teaching program in which the development of concepts and skills as well as the ability to apply mathematics, was pursued by means of mathematical modelling focused around a sequence of carefully selected problems. The program was implemented within the normal school organisational framework, and thus inherited classroom structures and student groupings defined by school policy. The college's Year 8 enrolment comprised three mathematics classes that were set on the basis of a Test of Learning Aptitude (TOLA), which included an assessment of the mathematical abilities of the entering students who came from a range of feeder primary schools. Three classes (defined as High, Medium and Low ability) were defined on the basis of the TOLA scores and the medium ability class (N=23) formed the trial group for the study. Both the trial group and the high ability group studied the same content, which as a typical first year secondary course contained a blend of simple algebra, geometry and arithmetic. The non-trial high ability group (N=23) was taught by traditional methods involving exposition of ideas, techniques, and worked examples by the teacher followed by consolidation exercises, undertaken by the students. For purposes of discussion this class will be referred to as the *conventional* group to indicate that it received mathematics instruction along conventional lines. The research set out to explore questions with respect to the impact of a modelling methodology on educational outcomes. Specifically to investigate the extent to which the program would:
1. impact on attitudes to mathematics;
2. generate understanding of associated mathematical concepts and skills;

3. engender improved competencies in processes of applying
    mathematics;
4. improve the quality and consistency of mathematical communication.

   The approach to the study is most appropriately located within the
case-study genre. It should not be thought of in terms of a formal
experiment, involving treatment and control groups - school constraints
precluded this in any case. However the relative performance of the two
groups on some measures was possible to obtain, and is of interest.

3. The Trial program (Mathematical Modelling)
3.1   The Modelling Process
   A six-step version of the commonly depicted modelling process
was taken as the basic anchoring framework for mathematical
development, which was implemented in the classroom of the first named
author.
1. Clearly specify the problem
2. List all data needed and assumptions made
3. Construct (or recognise) the model to be used
4. Perform any necessary mathematics (to form the model or use a
    known model to solve a problem)
5. Check the solution (verify, validate), use the solution to predict, and
    make necessary changes
6. report the results, detailing all steps

   While such frameworks have been used for many years within
modelling programs, there appears to have been a missing link between
the use of such heuristic devices and pedagogical theory. For example,
many of the teaching illustrations discussed at ICTMA conferences have
effectively used a version of this modelling process, apparently because it
has proved useful or 'it works'. We believe that this mathematical
tradition can be linked with educational principles, situated within a
sociocultural orientation to learning, which aims to develop the connection
between classroom activity and that of a mathematical community of
practice. Such an approach is supported by Vygotskian concepts of
scaffolded learning within zones of proximal development (ZPD),
(Vygotsky, 1978). The scaffolding notion of ZPD (Vygotsky, 1978;
Palincsar, 1986) refers to the development of individual expertise through
appropriate guidance from more capable others, including peers or

teacher. So here we elevate the role of the modelling framework beyond the summary of mathematical process, to that of an essential participant in the learning process within which it provides metacognitive scaffolding to assist the development and monitoring of associated mathematical learning. Another scaffolding aid was provided in the form of the criteria-standards grid (Figure 1).

Ability to specify problem clearly	Ability to set up model	Ability to solve maths, interpret, validate, refine	Ability to communicate results	
Can proceed only when clues are given	Can proceed only when clues are given	Can solve maths problem given assistance through clues and hints	Communicates reasonably re: • presentation • conciseness • oral reporting	C
Can extract clues from information and clearly define the problem	Can develop model with minimum assistance	Can solve the basic problem with no assistance	Communicates clearly in the above categories	B
Can extract clues, define problem if information is open-ended, insufficient, redundant	Can develop model independently where no clues exist	Can solve the problem independently. Can evaluate and refine the model	Communicates clearly in the above categories using innovative and creative ideas	A

Figure 1:Criteria with Standards for Assessing Modelling Performance

This grid, which links closely to the modelling process, served two cumulative purposes throughout the program. In providing information on performance on four modelling criteria, at three different standards of quality, it acts as a formative assessment procedure, through which immediate detailed structured feedback is provided for students on their solution attempts. Secondly, through this feedback students receive an indication of the target quality to be achieved. For each of the 12 cells information is received, assessing the present effort (A, B or C) together with verbal comment, against the standard expected (A). This grid therefore acts as another structure (external referent) to scaffold student learning, as well as a public basis for assessing the summative quality achieved.

3.2   *Group activity.*

The 'C' in Collaborative Learning has been used somewhat ambiguously to refer to both cooperative-based learning (group members share the workload); and collaboration-based learning (group members develop shared meanings about their work), (Webb and Palincsar, 1996). While interrelated there is a clear difference in the respective emphases. Collaborative activity, in this latter sense, is characterised by equal partners working jointly towards an end (Anderson, Mayes, and Kibby, 1995), and this was the emphasis deemed appropriate within the modelling context. Collaborative activity in this sense is supported by another interpretation of the ZPD - its application in egalitarian partnerships. This assumes there is learning potential in peer groups where students have incomplete but relatively equal (mathematical) expertise - each partner possessing some knowledge and skill but requiring the others' contribution in order to make progress. Here, it is important to recognise that not all student constructions are equally valid, although all are equally legitimate as a basis from which to proceed to greater understanding. Hence the teacher, as a more experienced knower in the discipline, has an essential role to play in selecting student ideas that are fruitful to pursue while passing over others, and in asking students to justify their conjectures, strategies, and solutions.

3.3   *Training Problems*

Introductory problems were designed to develop the students' understanding of the modelling process. These problems involved familiar, life-related situations and used well-established mathematical knowledge. The first example used in the program, required students to collect much of their own data and make appropriate assumptions in using it.

> *A bus trip to (or from) school costs 50 cents and John's parents are considering the more economical option - to buy John a bike to ride to school or for John to catch the bus? The students were expected to provide a mathematical argument for or against the purchase of a bike, that took into account all relevant costs.*

This first problem took more than one week to complete, as the students suggested and included various costs such as replacement tyres, insurance, helmets, raincoats and repairs. They discussed what variables could be included in a *costing* model, eventually realising that factors like safety and enjoyment could be included in a report, but were difficult to evaluate for inclusion in a model. The students recognised that different options needed to be provided for in developing the costing, such as the possibility of increased bus fares during John's period at school, or John shifting house. Most considered at least two different priced bikes, showing the bus option as better for one bike, while not for another, and there were solutions that considered different numbers of years that John may attend school. The students wrote individual versions of their models.

3.4   Mainstream problem-based development

Following the completion of the training problems the students embarked upon the remainder of the program which extended over the rest of the year. Mathematical concepts and skills were introduced and developed through application contexts presented as modelling problems. Simultaneously the process of modelling itself was practised and re-inforced, as students used the structure of the six-stage modelling process. The Appendix contains an example of a resource problem provided for this purpose.

3.5   Other teaching procedures

The students had their acquired skills regularly reviewed and reinforced within later application problems. However they were provided with additional experience in that traditional homework exercises were set from a standard text after they had been introduced through practical contexts. On the one hand this provided concept reinforcement and skill practice after the particular content had been given meaning in an applied situation, while on the other it provided a measure of comfort and familiarity for students feeling anxiety in adjusting to a teaching/learning approach different from the one they were used to. The students were also expected to write journal entries following each session. These journals used *expressive writing* to reflect to the teacher their feelings about their work, what was enjoyable, what caused difficulties, and how they felt their group was operating.

4. Assessment measures

Students in the trial were required to participate in assessment exercises as a necessary part of normal school reporting procedures. Some of these (tests of recall, basic skills and content) were common to both the trial and conventional groups. Other assessments were specific to the trial group - in particular those which set out to obtain evidence of modelling and applications competence. Class assessment for the trial group included:

(a) a range of modelling problems set throughout the year that were completed in class or home-time;
(b) a supervised extended "modelling problem" in semester 2 conducted in a class-time 'test' situation;
(c) a traditional end of year examination testing 'recall and access' of the total course content (taken also by the conventional group).

5. Data Sources

Performance data were collected regularly, and student observations conducted on a daily basis. The teacher/researcher maintained a reflective diary throughout the year in which observations of student activity and behaviour were recorded. This augmented and cross-checked with the material in the student portfolios (that included modelling problems, written performance data, and diary entries), and audiotapes of oral testing. Additional data sources were built into the program as discussed below.

An attitude survey was developed from an instrument (Aiken, 1979) with four sub-scales relating to *Enjoyment (E), Motivation (M), Importance (I),* and *Freedom from Fear (F)*, being represented by Likert-scale items on the usual strongly agree to strongly disagree axis. Pre (February) and post (November) data were obtained for both the trial and conventional groups to identify attitude shifts that might be related to the teaching programs. The range of scores for each six-item sub-scale varied from 6 to 18 with 6 representing the most positive response, so that totals could vary from 24 to 72.

In summary, data collection involved the following quantitative and qualitative forms.

- Quantitative data comprised entry (TOLA) aptitude scores, scores on traditional test instruments and individual (pre and post) scores on the structured attitude survey.
- Qualitative data included evaluations of modelling performance using the categories in Figure 1, excerpts from students' journals; records of oral interviews with students; and excerpts from the teacher's diary. Additionally, selected students were interviewed in the September of the year following the trial.

6. Outcomes

Although a case study approach was adopted, some comparative data analysis has been deemed illuminating, although not in the sense of experimental versus control group design. With respect to both the trial and conventional groups, the analysis of pre and post attitude response data is useful for purposes of assessing the impact of the respective programs on students' feelings in relation to mathematics. Also, because the school's reporting policy meant that a summative test instrument was common to the groups, it was useful to compare performance on this instrument. The purpose here is descriptive not inferential - that is it is worth asking whether, for these groups of students, differences in performance are enough to suggest that something with an interesting educational impact is occurring.

6.1   Attitude data

Very little shift occurred in the means of either group. Considering total scores, pre-post means for the trial group were 36.7 and 35.8, and for the conventional group 30.8 and 31.1. There was more variation (both pre and post) within the trial group: standard deviation 9.7 and 11.2, compared with the conventional group 6.2 and 7.7. For the trial group, 10 students became more negative (increased total score) and 12 more positive about their overall attitude to mathematics, while the corresponding comparison group movements were 11 and 9. This seems thoroughly unremarkable. However closer observation indicated that there were 10 shifts of more than 5 units pre-post in the trial group compared with only 5 shifts of more than 5 units in the comparison group. It appears that the students in the trial were affected more decisively (one way or the other) than those in the group taught by conventional methods.

6.2   Year 8 Mathematical Performance

As indicated previously the trial and conventional groups engaged in some common assessment involving a test of basic content and skill knowledge *(c)* administered at the end of the year. The material relevant to this test was covered in both groups albeit through different teaching approaches. Means (standard deviations) were for the trial group 77.7 (17.9), and for the conventional group 64.0 (19.0) with N=23 for each. An ANOVA found significantly in favour of the trial group: $F (1, 44) = 6.38$, $p < .05$.

The trial group also engaged in tailored assessment that provided data on the development of the problem solving and modelling skills of the students in the experimental program. For the modelling problems *(a)*, and the modelling test *(b)*, assessment was by means of the letter grades defined in Figure. 1, with the final award being determined from the total profile of individual letter grades achieved by each student. The range of such grades covered each of the 10 levels from A+ to D (no progress) where C+, B- etc indicate borderline judgments in applying the modelling criteria. The following excerpt is from the modelling report of one of the better students.

The modelling project was expressed as follows: You are an industrial designer who is trying to find the best specifications (measurements) of a cylindrical can which will hold exactly 750 ml. What do you advise?

When I started this model I didn't know what size the cans would be and what they would be used for. After I had made up my table and put measurements together to get 4 cylinders each containing just over 750 ml, I cut out models from paper. This gave me an idea of what I could use the cans for. Can 1, which is 26.6 cm tall and 6 cm in diameter, could be used for a Thermos. It is the right height and is a good width so that you can grip onto it. Its capacity is 752 ml. The surface area is 558 cm^2. It would be the dearest size shape to make because it has the largest surface area. The reason I have the capacity a little larger than 750 ml is so there is room between the liquid and the lid. Can 2 would be a pretty good size beer mug for a big drinker. The height is 15 cm and the diameter is 8 cm. It's capacity is 754 ml and surface area is 477cm^2. Again the can is not full to the brim. The can, which will be cheapest to make, is can 3. It has a surface area of 459cm^2 and is 9.6 cm high and 10 cm in diameter. It could be used for canned juice.

6.3   Year 9 (common assessment)

In the following year, the trial and conventional group members were reconstituted into year 9 classes that were taught by conventional methods viz. exposition of ideas and techniques by the teacher, worked examples by the teacher, followed by worked exercises by the students. While the absolute performance differential in favour of the trial group students disappeared on the year 9 traditional test (mean 74.2 versus 82.3), when TOLA scores were entered as a covariate, their adjusted performance remained ahead of the comparison group, $F(1, 36) = 4.71$, $p < .05$. So, allowing for differences in entering ability, the trial students continued to achieve soundly on traditional measures relative to their higher ability counterparts in the conventional year 8 program.

6.4   Student reactions

Five target students were selected as representative of the variety of ability and attitude present in the trial class; one subsequently left the school. Observational data were collected intensively for these students, and we include below excerpts from transcripts of interviews that took place in the September of the year following the trial. This represented a 10 - month time lapse since the conclusion of the year 8 academic year. The year 9 interviews were structured around selected questions (see sample questions 1 - 4 below) with opportunity provided for elaboration and initiative in student responses.

1. Did you notice any advantages or disadvantages in having done mathematical modelling last year?
2. Have you used the 6 steps of modelling again?
3. Do you miss group work?
4. Have you any other comments about last year?

John (a student with the third most negative overall attitude rating in November of year 8).
1. Modelling was easier than textbook work.
2. Our problems this year have been too easy to need the six steps. You know! Things like – if 3 pens cost so much, what does one pen cost?
3. At times yes, but I stuffed around a lot in groups. I'm probably better on my own. I miss some of the activities.
4. I'm glad we did it 'cause I'm doing good this year.

Taylor (a student with an ambivalent to negative attitude throughout the trial)

1. Yes, we learnt a bit more than the other students most of the time. But if I am not really listening at times I miss what is being taught. The teacher goes too quick and I can't catch up like I did last year when the group helped me understand what I missed.
2. No. I would rather do modelling though, than boring textbook work. (Teacher: You continually complained about modelling all last year. "When are we getting off this?" "Why do we have to write so much in maths?") Yes, but it was much better than this year.
3. It is better in a group. You can get help and work at own speed.
4. No. Just that it was better than now!

Graham (a student whose attitude improved substantially over the trial program)

1. I enjoyed modelling last year and keeping a notebook helped me at the end of semester exams because I could revise what I had learnt.
2. I've thought about setting out, but not the actual steps.
3. No. I would rather work in groups, than at separate desks. It took a while to get used to working on my own. I miss group help.
4. I would enjoy to continue modelling next year because I am not enjoying the things we are doing now.

Gary (a positive motivated student)

1. It did allow us to learn new concepts and it helped us understand them easier. It improved our setting out and communication.
2. No. I am careful with setting out though.
3. I am happy to work on my own, but I guess I miss hearing other kids' ideas.
4. I really would rather be doing modelling.

7. Reflection on Outcomes

The pre-post attitude data as viewed from either end of the year 8 experience do not indicate strong effects. However interview data from the target students during year 9 adds another dimension suggesting further study. All four students were retrospectively positive about the modelling program (question 4) with three of them making negative comparisons regarding their current year 9 course. Two of these students had been negative towards the year 8 program, one of them extremely so. This raises the question of the post-modelling experience contributing a dimension important for the evaluation of the modelling program itself - that reflection following a return to a traditional program may modify immediate post-program reactions? For example whether such a change in classroom culture is viewed at the time simply as more 'mathematics stuff', particularly by those not kindly pre-disposed towards the subject? If so, the impact on attitudes of an innovative program may be difficult to assess immediately, with later reflection possibly an important component of a reliable evaluation.

With respect to performance, while the trial group students developed skills in application and modelling they also significantly outperformed the conventional (high ability) group on the summative assessment task common to the two groups in year 8. In contrast to the trial group students, those in the conventional group were consistently penalised for poor setting out and for failing to include and explain necessary working, although instructed explicitly to do so. The teacher of the conventional group insisted that his students were made aware of such requirements, which had been stressed throughout the year, and indeed the high ability students themselves assured their teacher that they felt well prepared for the exam and were "surprised by the results". We appear to have here an example of the difference between 'knowing that' and 'knowing how'. The conventional students 'knew that' they were required to communicate reasons and explanations, but had not made this a habit. By contrast the modelling students 'knew how' to do it, having developed the ability as an integral part of report writing. In year 9 the absolute performance differential in favour of the trial students was reversed, although when entry ability was controlled for, their relative performance compared favourably with the students from the high ability year 8 class.

It remains to explore aspects of the year 8 modelling program that might explain the relative performances in both years 8 and year 9. Fundamental to the approach was the structured modelling procedure (six-step version) that scaffolded the students' approach to their mathematical learning. The criteria-standards grid complemented this contribution by providing both explicit targets and a basis for the provision of continuous feedback on the degree of attainment of these targets. The structured modelling procedure acted effectively as a series of external metacognitive prompts, which helped to guide the students' learning as they worked their way through a careful selection of problems. Arguably this promoted consistency in attacking mathematical questions including particular attention to explanation and written communication. The selected year 9 student interviews indicated that this structured approach was not continued (at least consciously) in the following year and indeed the trial students appeared to suggest that the "problems" encountered there did not require it! An explanation for the change in relative performance of the trial students and the conventional students between years 8 and 9 may be conjectured on this basis. That is, that while the trial students adjusted smoothly enough to a change in teaching style they also abandoned the key element that advantaged their learning in the previous year. Put another way the modelling approach was not internalised by these junior school students, but remained an external scaffold and as such was vulnerable to changed classroom conditions. At the same time there appeared to be no impediment to the students moving from the modelling based context to conventional teaching for they continued to achieve well on common assessment in relation to their mathematical abilities. And in this respect they were also constrained to play on "opposition turf" for there was no opportunity in their year 9 program to display through assessment their capacity on modelling type activities aligned to their year 8 program.

Finally it is appropriate to reflect on the total program. At the outset of a project such as this certain expectations are held, some of which are realised and some of which are not. Additionally, unforeseen elements emerge, and it becomes a challenge to relate the totality to dimensions which link with the wider world of existing theory, or suggest directions for its extension. This contrasts with experimental designs, which set out to test pre-determined hypotheses. In this case the project had elements of a voyage of discovery, including outcomes that could not have been reasonably predicted. What has been created in addition to student outcomes, is a methodology that is capable of reproduction, and hence of providing a means of testing or challenging the robustness of the outcomes that this experimental program of 'learning through modelling' has produced.

References

Aiken L (1979) 'Attitudes towards mathematics and science in Iranian schools' *School Science and Mathematics* 79, 229-234.

Anderson A, Mayes J T and Kibby M R (1995) 'Small group collaborative discovery learning from hypertext' in O'Malley C (Ed.) *Computer Supported Collaborative Learning* New York: Springer-Verlag, 23-38.

Blum W and Niss M (1991) 'Applied mathematical problem solving, modelling, applications and links to other subjects - State, trends and issues in mathematics instruction' *Educational Studies in Mathematics* 22, 37-68.

Palincsar A S (1986) 'The role of dialogue in providing scaffolded instruction' *Educational Psychologist* 21, 73-98.

Vygotsky L D (1978) 'Mind in Society: The Development of Higher Psychological Processes', Cambridge MA: Harvard University Press.

Webb N M and Palincsar A S (1996) 'Group processes in the classroom' in Berliner D C and Caffee R (Eds) *Handbook of Educational Psychology* New York: Macmillan, 841-873.

Appendix: Example of a 'Teaching Project' (*Dog Trials*)

At a country show, sheepdogs have a specified time to muster as many sheep as possible into a pen. Each dog was given one practice run before the final trial. The number of sheep herded for each of 15 dogs is:

Dog	Practice (no. sheep)	Final (no. sheep)	Dog	Practice (no. sheep)	Final (no. sheep)
1	40	59	9	31	49
2	62	63	10	86	98
3	71	81	11	75	80
4	24	40	12	76	67
5	45	55	13	42	35
6	22	29	14	51	51
7	39	43	15	59	70
8	58	72			

Problem. You are asked by the judges of these trials to use these results to find a general model that could be used to predict results in finals from results in practice.

Activity.
1. Graph the results of FINAL against PRACTICE.
2. Estimate and draw in what you think is the LINE OF BEST FIT i.e. keep distances above the line the same as below.
3. If 50 sheep were herded by a dog in practice, what is a reasonable estimate for the number herded in the final?
4. If a dog herded 60 sheep in a final, what did it probably herd in practice.
5. Use the computer to get the graph and the line of best-fit model.
6. Use your model to re-answer questions 3 and 4.
7. Write your report for the judges.

3

Using Ideas from Physics in Teaching Mathematical Proofs

Gila Hanna
Ontario Institute for Studies in Education of the University of Toronto
gila_hanna@tednet.oise.utoronto.ca

Hans Niels Jahnke
Universität Essen, Germany
njahnke@uni-essen.de

Abstract

An important challenge faced by mathematics educators is to find effective ways of using proof in the classroom to promote mathematical understanding. A very promising approach that has been insufficiently explored is investigated in this paper: The use of arguments from physics within mathematical proofs in the context of teaching the lever principle and the principle of the unique centre of gravity of a triangle as premises in mathematical proof.

1. Introduction

This paper advocates the investigation of one very promising approach to teaching proof that has been insufficiently explored namely the use of arguments from physics within mathematical proofs. The use of physics under investigation goes well beyond the mere physical representation of mathematical concepts. What is examined is the classroom use of proofs in which a principle of physics, such as the uniqueness of the centre of gravity, plays an integral role in a proof by being treated as if it were an axiom or a theorem of mathematics. This use of physics is also entirely distinct from "experimental mathematics", which purports to employ empirical methods to draw valid general mathematical conclusions from the exploration of a large number of instances.

The paper also describes empirical research into the effectiveness of using concepts and principles of physics in teaching geometrical proofs. Motivating this research was the expectation that drawing upon physics in the mathematics classroom would promote understanding, since proofs that invoke physical arguments appeal more directly to our physical experience and are often more easily grasped in their entirety, and are thus potentially more easily understood and more readily remembered by students. It was anticipated that proofs using concepts and principles of physics would support the goal of not only proving that a mathematical proposition is true, but also showing clearly why it is true. This approach is consistent with the significant reorientation in the thinking of many mathematics educators, especially over the past twenty years, towards using intuitive appeal in the teaching of proof.

While applications of mathematics to physics are well known and common, applications of ideas and laws of physics to the solutions of mathematical problems or to the teaching of proof are not as well known (Hanna and Jahnke, 1993, 1996, 1999). Nevertheless, several mathematics educators and mathematics organisations have advocated basing the teaching of mathematics upon its various applications. The Freudenthal Institute has developed a theoretical framework which promotes the use of realistic problems as a source for mathematization (Freudenthal, 1983, Streefland, 1991), a framework now referred to as Realistic Mathematics Education (RME). The Organisation for Economic Cooperation and Development (OECD) has also recommended a closer relationship between mathematics and the natural sciences (OECD, 1991) and has supported projects that in various ways seek to strengthen the role of applications in mathematics teaching. The International Conference on the Teaching of Mathematical Modelling and Applications (ICTMA) has published many papers demonstrating the value of modelling and real-world applications in the teaching of mathematics (see Niss et al, 1991, Blum et al, 1989). A rich source of examples of the application of mechanics to mathematics can be found in Uspenskii (1961), in Winter (1978, 1983) and in Polya (1954); none of these applications deals explicitly with the teaching of proof, however.

The use of arguments from physics in mathematical proofs may be seen as part of a broader trend of constructing special contexts or media of argumentation, mostly of a geometric nature. These contexts have been found to be more accessible to students since they have an advantage over purely algebraic or symbolic proofs, in that they may possibly assist the student in understanding the necessity of a conclusion. As examples of contexts that were found to be effective in the teaching of proof, we cite the classroom application of the concept of preformal proofs (Semadeni, 1984, Blum and Kirsch, 1991), the judicious use of the concept of "intuitive proof", more appropriately termed *inhaltlich-anschaulicher Beweis* in German, (Wittmann and Müller, 1988), and the distinction between explanatory and non-explanatory proofs (Hanna, 1990, Dreyfus and Hadas, 1996).

2. Applications of laws of physics to mathematical proofs

The proposed use of concepts and principles of physics goes well beyond the mere physical representation of mathematical concepts, such as the rectangular grid (array of m by n points) that is often used to illustrate the concept of commutativity in multiplication (that is, that m x n is the same as n x m). Such a grid does make it clear that the order of multiplication does not matter, but it is no more than a visual representation. What is advocated by Hanna and Jahnke (1999), among others, is the classroom use of proofs in which a principle of physics, such as the uniqueness of the centre of gravity, plays an integral role, either by forming a premise or by being treated as if it were a theorem of mathematics.

As an example of the application of laws of physics to mathematical proofs, we might first mention the construction of the Fermat point of a triangle. The Fermat point of a triangle ABC is the point X which yields the minimal sum of distances from X to the three vertices (Figure 1).

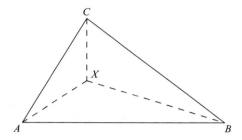

Figure 1: Optional location of a school

An example would be a school serving three cities A, B, and C, located so that the sum of the distances from the school to the cities is minimal. The most elegant way to find the Fermat point is to model the triangle by a physical system consisting of a perforated plate and weighted ropes, as shown in Figure 2 (Polya, 1981).

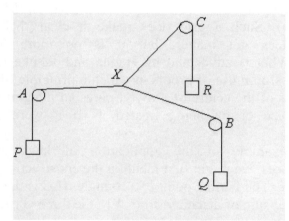

Figure 2: Mechanical device to find the Fermat point

Two different mechanical principles feature in this example: first, the addition of forces by the parallelogram rule and, second, the fact that a system of connected bodies in a gravitational field tends to a state where its common centre of gravity is in its lowest possible position (i.e., where the potential energy is minimised).

To which position will X move? Of course this must be a position where the three forces acting on X due to the weights balance each other. In addition, X must move to the point that creates the state of minimum potential energy. This is the point at which the three weights P, Q, R, taken together, hang as low as possible. This means that $AP + BQ + CR$ takes on a maximum. Since the sum of the lengths of the three strings is constant, it follows that X will move to the point at which $AX + BX + CX$ takes on a minimum. In other words, X will move to the Fermat point of the triangle ABC. Since in the position of equilibrium the three forces balance each other and since the weights are equal, then, for reasons of symmetry, the directions of the forces must be equally inclined to each other. Therefore

$$\angle AXB = \angle CXA = \angle BXC = 120°$$

The required point X is easily constructed.

This proof, relying upon a mechanical rather than mathematical principle, is nevertheless an insightful and convincing mathematical proof. The Fermat point is characterised by a sort of symmetry, and the physical analogy makes the nature of this symmetry immediately clear.

2. A Teaching Experiment

Another example of the use of principles from physics in mathematical proof is the use of the following three postulates of statics:

Postulate 1: In any system of masses, the position of the centre of gravity of the system depends on the on the position of the masses; in particular the centre of gravity of a system of two masses lies on the straight line joining these masses and its distance from each mass is inversely proportional to that mass.

Postulate 2: Any system of masses has only one centre of gravity.

Postulate 3: In any system of masses, if any two individual masses are replaced by a single mass equal to the sum of the two masses and positioned at their centre of gravity, the location of the centre of gravity of the total system remains unchanged.

These three postulates were used in a teaching experiment in which ideas from physics were introduced into the teaching of proofs in geometry. The purpose of the experiment was to examine in what ways arguments from physics might help students understand a proof, whether they were superior to geometric arguments in creating understanding, and whether their use in the teaching of proof could be recommended as a promising pedagogical approach.

The specific task undertaken was to prove that the medians of a triangle are concurrent at a point of trisection. The two objectives of this teaching experiment were 1) to have the students discover that the medians of a triangle intersect at a single point (the centre of gravity of the triangle when three equal masses are placed at its vertices) and that this point is located on a median, two-thirds of the way from each vertex, and 2) to have the students provide a proof or explanation why this must be so, using the three postulates related to the centre of gravity of a system.

To facilitate this experiment, a new teaching unit was introduced into a grade 12 mathematics course held in the second semester of the school year, in Toronto, Canada. The course was attended by 29 students, 15 males and 14 females. The median age of the students was 16.5, with a minimum age of 15 and a maximum of 18. Two classroom periods of 75 minutes each were allocated to the unit.

The students were not taking a course in physics concurrently and were not expected to be familiar with principle of statics. Therefore, some teaching of statics was incorporated into the mathematics class. The first period introduced the lever principle formulated as follows: A balance beam or a lever is in equilibrium whenever the product of the mass and the arm length on one side of the fulcrum is equal to the product of the mass and arm length on the other side (Figure 3).

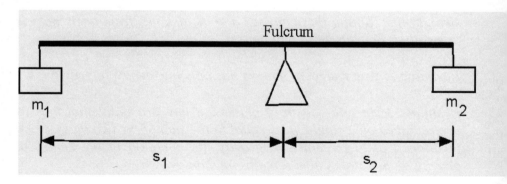

Figure 3: Balancing masses

This lesson required the students to work in independent groups of three with a minimum of guidance from the teacher. Each group was provided with a clear acrylic triangle, a washable marker, a ruler and a number of 200g masses. Each student was also provided with a set of four worksheets for this experiment. The triangles were of varying shapes and sizes, including acute, obtuse and right triangles. Each triangle had six string loops taped to the corners of the triangles and at the midpoints of the sides. The students were told to follow the instructions in the worksheets. These instructions required the students to find the balancing point with the masses suspended from the vertices of the triangle, and then to verify that this point remained unchanged when two masses were relocated from their respective vertices to the midpoint of the side joining the vertices where the masses came from (Figure 4). Clearly, this exercise required a good deal of experimentation and trial and error. The students were also taught how to prove the theorem using only geometric arguments.

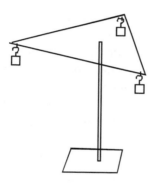

Figure 4: Balancing a triangle on the retort rod

During the second period, all the participants were asked to write out complete proofs of the theorem about the medians in a triangle in two ways, one geometric and the other using arguments from physics. In the latter case they were to apply the above three postulates and explain how the experiment of balancing a triangle proves that the medians of a triangle all intersect at one point.

3. Results and discussion

The responses and comments indicated that the idea of using arguments from physics has a great deal of appeal for students. An important specific finding of this study is that in response to the question "Is the argument from physics convincing?" all but two of the 25 participants (out of 29) who responded to the questionnaire found the proof using the argument from physics convincing. Responses to two other questions yielded the following results:

| Which proof is clearer? | Physics (15) | Geometry (9) | Don't know (1) |
| Which is easier to remember? | Physics (15) | Geometry (5) | Don't know (5) |

The interview responses give further insight into the issue of "convincing." One student remarked that "the argument from physics actually made sense," and another responded that in using physics "there is a hands-on, 3-D, example of why the argument is true." Yet another student remarked that "it is difficult to see how the physics argument is related to the issue at hand earlier on, but the connection eventually becomes more evident and concrete than a geometric proof could ever be." (For additional details, see Hanna et al, 2000.)

The students' proofs, the geometrical and the physical, were assigned by the classroom teacher to one of three categories: Correct response, partially correct response (including incomplete responses with a correct start or with a correct use of arguments) and incorrect response (including responses with minor correct parts but mostly wrong). The results indicate a much higher rate of success on the purely geometric proof, with 23 of the 25 students producing a correct proof. Only 12 students produced a correct proof using arguments from physics, while 9 produced a partially correct proof and 4 an incorrect one.

Though these results are disappointing, in that they did not lend support to the didactic value of physical principles in the teaching of proof, it seems fair to assume that they reflected previous classroom work, in which the students had become quite accustomed to the geometric approach. Indeed, the participating students had never had any previous instruction in the use of physics, but had some traditional geometry (angle theorems and congruence postulates) in Grade 10.

From the students' comments and their responses to questionnaire questions, it seems quite clear that many students felt that the experiments were part of the proof. These rudimentary experiments had been introduced into the teaching unit to establish in the students' minds the empirical plausibility of the physics principles used in the proof as well as of the conclusion of that proof. In this the experiments would seem to have been successful. They also appear to have been successful in conveying to the students the more general idea that concepts and principles from physics can be used in proving mathematical theorems. But unfortunately the experiments may also have misled students into viewing a proof based upon physical concepts and principles as in essence a generalization from empirical observation. This potential pitfall will have to be taken into consideration in any further exploration of the use of concepts and principles of physics in the mathematics classroom.

References:

Blum W, Berry JS, Biehler R, Huntley I, Kaoser-Messmer G and Profke L (Eds) (1989) 'Applications and modelling in learning and teaching mathematics' Chichester: Ellis Horwood Limited.

Blum W and Kirsch A (1991) 'Preformal Proving: Examples and Reflections' *Educational Studies in Mathematics* 22, 183 - 203.

Freudenthal H (1983) 'Didactical phenomenology of mathematical structures' Dordrecht: Reidel.

Dreyfus T and Hadas N (1996) 'Proof as an answer to the question why' *Zentralblatt für Didaktik der Mathematik* 28, 1, 1-5.

Hanna G (1990) 'Some pedagogical aspects of proof' *Interchange* 21(1), 6-13.

Hanna G and Jahnke HN (1993) 'Proof and application' *Educational Studies in Mathematics* 24, 4, 421-439.

Hanna G and Jahnke HN (1996) 'Proof and proving' in Bishop A et al (Eds.) *International Handbook of Mathematics Education*, Dordrecht: Kluwer Academic Publishers, pp. 877-908.

Hanna G and Jahnke HN (1999) 'Using arguments from physics to promote understanding of mathematical proofs' in Zaslavsky O (Ed) *Proceedings of the twenty-third conference of the international group for the psychology of mathematics education*, Haifa, Israel: Technion Printing Centre, 3, 73-80.

Hanna G, Jahnke HN, DeBruyn Y and Lomas D (2001) 'Teaching mathematical proofs that rely on ideas from physics' *Canadian Journal of Science, Mathematics and Technology Education*, 1, 2, 183-192.

Niss M, Blum W and Huntley I (Eds) (1991) 'Teaching of Mathematical Modelling and Applications' Chichester: Ellis Horwood.

OECD (1991) 'Proceedings of the second conference of the OECD project: Future perspectives of mathematics, science and technology education in OECD countries', Paris, France: OECD

Polya G (1954) 'Mathematics and plausible reasoning. Vol.1: Induction and analogy in mathematics', Princeton: Princeton University Press.

Polya G (1981) 'Mathematical discovery: On understanding, learning and teaching problem solving' (2 volumes combined.), New York: John Wiley and Sons.

Semadeni Z (1984) 'Action proofs in primary mathematics teaching and in teacher training' *For the Learning of Mathematics* 4 (1), 32 - 34.

Streefland L (1991) 'Fractions in realistic mathematics education: A paradigm of developmental research', Dordrecht: Kluwer Academic Publishers.

Uspenskii VA (1961) 'Some applications of mechanics to mathematics' (translated from the Russian by Halina Moss) New York: Blaisdell Publishing Company.

Winter H (1978) 'Geometrie vom Hebelgesetz aus - ein Beitrag zur Integration von Physik- und Mathematikunterricht der Sekundarstufe I' *Der Mathematikunterricht* 24, 5, 88-125.

Winter H (1983) 'Zur Problematik des Beweisbedürfnisses' *Journal für Mathematik-Didaktik* 1, 59 - 95.

Wittmann EC. and Müller GN. (1988) 'Wann ist ein Beweis ein Beweis?' in: Bender P (Ed) *Mathematik-Didaktik: Theorie und Praxis, Festschrift für Heinrich Winter*, Berlin: Cornelsen, 237 - 257.

4

Deconstructing Mathematical Modelling: Approaches to Problem Solving

Christopher Haines
City University, London, United Kingdom
c.r.haines@city.ac.uk

Rosalind Crouch
University of Hertfordshire, Hatfield, United Kingdom
r.m.crouch@herts.ac.uk

Andrew Fitzharris
University of Hertfordshire, Hatfield, United Kingdom
a.fitzharris@herts.ac.uk

Abstract

Students' developing expertise in mathematical modelling is analysed through comparison between expert and novice problem solving behaviours. Using multiple-choice modelling questions, reflective student questionnaires and interviews, this research has categorised processes used in solving problems and established connections between modelling, mathematics and perceived interest.

1. Research context

Mathematical modelling in higher education and possibly in other sectors operates under two teaching and learning paradigms:

(A)   a holistic approach in which students learn through experience of complete case studies in mathematical modelling. These might of course be through modelling problems that admit simple and straightforward outcomes progressing through to more difficult situations as confidence and competence increases.

(B)   the teaching of mathematical modelling by defining and examining in detail the processes and stages though which the modeller passes.

Representations of the modelling cycle by the seven box diagram or stages drive approach (A), whilst (B) is more reflective in its use of such representations.

Our research has been concerned with (B), using methods and procedures that provide for a snapshot of students' modelling skills at key developmental stages. In a study of 28 students at two universities Haines et al (2000, 2001) highlighted difficulties that students face at the beginning of the modelling process:

A. in identifying broad assumptions influencing a simple model

B. in posing clarifying assumptions and making a related mathematical formulation.

At the end of the process they also experience difficulty in making an effective comparison of the model outcomes with the real world problem.

This research and a further study (Haines and Crouch, 2001) used multiple-choice questions as a research tool, 12 questions having been constructed in 6 analogue pairs. These were then appropriate for pre-test and post-test investigations of attainment in modelling. Haines and Crouch (2001), in a study of a further 42 students, showed that the 12 items contributed to a mathematical demand continuum, collectively exhibited consistent behaviours and within each analogue pair the two items performed in a comparable manner, except for minor variations when specifying the model (questions 5 and 6). The pairs could show attainment in moving from the real world to the mathematical model (questions 1 and 2) and vice versa (questions 11 and 12). Attainment in mathematizing and clarifying assumptions could be addressed by questions 3 and 4. Each pair within questions 5 -10 could address attainment in formulating a model. Table 2 locates each pair in the modelling cycle. This research did not address other aspects of the modelling process: solving mathematics, interpreting outcomes, refining a model, or reporting.

2. Novice and expert behaviours

It is helpful to consider what might be involved in developing from a novice to an expert mathematical modeller. Schoenfeld (1987) reports that expert mathematicians spent time analyzing geometry problems and would back-track when necessary unlike students, who tended to read the problem, immediately make a decision and stay on that path regardless. This accords with the observation of Galbraith and Stillman (1999) that the expert modeller continually returns to re-examine and to redefine the assumptions.

Sternberg (1997) reports that experts have more domain-specific knowledge than novices, and it is better interconnected. Novices' knowledge is more loosely organised with inadequate information about application and more difficulty in making connections to any other relevant information they may have. These are characteristics, which current practitioners recognise in students new to modelling. Experts have superior knowledge enabling them to apply underlying principles, whereas novices often focus on similarity of surface features given in the problem statement and try to produce equations to solve the problem, without any prior qualitative analysis (Chi et al, 1981).

Unlike novices, experts are more able to usefully infer from domain-specific information, because they can recognise relevant information and can combine information together in ways unexpected to novices (Sternberg, 1997). In physics and mathematics, novices tend to work backwards from the goal, but for more complex problems, or with more experience, might work forwards from the problem statement (Heyworth, 1999). However, though Priest and Lindsay (1992) found that both experts and novices tended used forward inferences when solving mechanics problems, novices were not very competent at it. Experts were much better at generating plans or principles before starting to solve a problem, whilst novices immediately started to generate formulae and equations (Heyworth, 1999). Whilst experts in domains such as mathematics, physics and medical diagnosis use forward reasoning successfully (and backward reasoning mainly for tying up loose ends), forward reasoning is very error-prone for novices (Patel et al., 1996 pp 132-3).

Expertise does take a relatively long time to acquire (Glaser, 1996), starting with a dependency phase with external support from a teacher, a transition phase and a final expert phase, so it is likely that novice modellers will have difficulties, particularly at the stages we have noted in previous research. We set out to investigate how novice and expert modellers approach various stages of the problem-solving process.

3. Research tools, method and application

We hypothesised that experts would keep the whole modelling problem in view whilst novices would choose a path and stay on it, even though it might not prove successful. Experts would look for an underlying model, possibly including a classification according to previous experience, whilst novices would concentrate on surface features. Novices would not readily identify with or be so interested in the modelling situation. Finally, in moving between the model and the real world, novices would find it harder to abstract from reality to create a mathematisable model in the first place. They would be more comfortable moving from the model to reality. We have also had a wider agenda of wishing to understand how students develop as mathematical modellers.

Having developed and tested 12 multiple choice questions situated in the mathematical modelling cycle (Haines et al 2000, 2001, Haines and Crouch, 2001), a further 4 questions explored graphical representations and connections between the real world and the mathematical world. In this study the following research tools were used:

  A.   16 multiple choice questions, developed in analogue pairs
  B.   A student reflective questionnaire on individual answers to 4 multiple choice questions none of which were analogues of each other
  C.   An interview by an experienced tutor following the completion of the 4 multiple choice questions and the reflective questionnaire

These were applied to 23 first year undergraduate students and four postgraduate research students at two universities. The undergraduates, who had completed five weeks' study of mathematical modelling, provided 92 multiple choice question responses so that each of the 16 questions was covered by them five or six times. The four research students each answered four different questions leading to a single 'expert' response for each question. These responses enabled the classification of approaches to problem solving and provided evidence to examine the consistency between experts (research students) and novices (first year undergraduates)

3.1 What identifiable processes are used?

In the reflective questionnaire (B), the respondents were asked:

Can you describe the process by which you came to choose this particular multiple-choice option and why? Describe...the steps you went through in making your decision in as much detail as possible, include changes of mind, backtracking, stops and starts etc. Please comment (as fully as you can) on each/some of the multiple-choice options that you did NOT choose.

These responses enabled the processes to be categorised: a, b or c (table 1), and in doing so, careful consideration was given to the location of the problem within the modelling cycle. The novices used processes (table 1) that successfully integrated the real world requirements with that of the mathematical model in 43% of the 92 cases; 40 responses were categorised (a). This link was absent in 35% of the cases; 32 responses were categorised (b). The remaining 22%, 20 responses, could not make effective connections. Table 2 clearly demonstrates differences in behaviours at various points of the modelling cycle.

label	descriptors of processes used
a	evidence of taking into account the relationship between the mathematical world and the real world input to the model
b	Some limited evidence of the above, such as (i) mentions having thought about the model, but little evidence that this has been done, or (ii) has obviously thought about the model, but lacks knowledge of the real world and/or mathematics to solve the problem effectively
c	(i) no evidence that the relationship between the mathematical world and the real world input to the model has been taken into account nor that a modelling perspective has been adopted or (ii) the problem has been looked at simply in real world terms, or entirely in terms of reasoning or maths (according its position in the modelling cycle) without reference to the needs of the model nor to the interface between the maths and the real world

Table 1   Categorising responses

	process			
questions	a	b	c	location in modelling cycle
1,2	0.18	0.18	**0.64**	real world to model
3,4	0.00	0.18	**0.82**	real world to model
5,6	0.00	0.33	**0.67**	specifying model
7,8	**0.58**	0.17	0.25	specifying variables
9,10	**0.91**	0.09	0.00	constructing equations
11,12	0.33	**0.58**	0.08	maths to real world
13,14	**0.91**	0.09	0.00	graphs to real world
15,16	**0.58**	0.08	0.33	using mathematics

Table 2 Students' responses classified
The proportion of responses classified a, b and c (n = 11 or 12).
Responses > 0.5 highlighted.

Linking real world requirements to the building of a mathematical model at the early stage of the modelling cycle presents considerable problems for novices. Up to 82% of the responses in this area (questions 1 to 6) fall in category (c), for example:

Q6 (supermarket checkout): novice 4, chose option E, full credit, process c

"I eliminated the others...because nobody can...predict demand and the number of items someone wants. I concluded this answer...by looking at...supermarket layouts. Tesco's...has about 30 tills and 4 express counters. This... reduces to option E. ... people with 8 items do not have to use express counters but people with large amounts of (items) do need to use the larger ones"

Although achieving full credit on this question, the student has tried to solve the problem rather than addressing the simulation and in doing so, he has used a means end analysis (Lesgold, 1988) drawing on his own real life experiences.

Q15 (load under hv cable): novice 22, chose option D, partial credit, process c

"I added the base height to the object height and ringed the... likely answer. I was (thinking about) using maths to find out but I (thought) I didn't need to as it 4 metres wide and if it (went) next to the pylon it would be able to get through without touching"

In not using the information given about the height of the hv cable she has relied on 'common sense'. This option does not allow the object to pass through.

These students do considerably better in areas where standard mechanical or interpretive procedures are used, establishing good modelling linkages. This can be seen in questions 9,10,13 and 14 where 91% of the responses fall in category (a) (Table 2). The expert responses all indicate processes in category a, (except for questions 11 and 12 where one response is coded a, and the other is coded b) moving easily between the real world problem and the mathematical model under investigation.

The difficulties that these novices had in the process of abstraction, in moving from the real world to the mathematical world, are confirmed in other disciplines. Patel and Ramoni (1997) studied medical students and expert doctors who, when making diagnoses, need to generate a hypothesis (including developing a model of the patient (Patel et al., 1996 p.131)). The first stage was to identify relevant problem features, ignoring those irrelevant to solving the problem (c.f. questions 1-6). Novices found abstraction difficult and used any available knowledge to solve parts of the problem. Experts, however, could quickly identify relevant problem features and choose a restricted set of hypotheses to explain them. Similar behaviour in solving geometry problems is reported by Koedinger and Anderson (1990).

3.2 How are distractors discarded?

Experts think first, analysing the problem and thinking about alternatives, whereas novices immediately embark on a course of action (Glaser and Chi, 1988; Priest and Lindsay, 1992; Schoenfeld, 1987). Reassuringly, in our study, novices showed some elements of expert behaviour. For example, of 90 responses almost two-thirds (59) gained credit (1 or 2 marks); of those gaining full credit (2 marks), the majority either straightaway discarded a zero credit (0 marks) option or did not discard any options immediately. Of the 16 expert responses, 12 discarded no options straightaway. 8 of these gained full credit and two further gained partial credit (1 mark). The remaining 4 responses discarded options credit zero, with each final choice gaining credit.

4. Case Studies

We now consider the novice responses and the expert responses, comparing and contrasting behaviours, on modelling questions 5 and 6 (specifying the model), 11 and 12 (mathematics to the real world) and 15 and 16 (using mathematics) [see Note 1]. A FACETS analysis (Linacre, 1990) of the novice data set suggests that: of questions 5 and 6, the preferred option (2 marks) is more likely to be selected for question 6, but, for both questions more than half will obtain a partial credit (1 mark). Of questions 11 and 12, students are more likely than not to obtain partial credit for question 11, but are unlikely to gain any credit for question 12. For questions 15 and 16, which require mechanical and interpretive procedures common in the mathematics curriculum, students are more likely to achieve full credit than partial credit. The expert responses to all these questions gained full credit. This analysis places our discussion of *interest* and perception of *what is mathematics* in the context of achievement in modelling.

Experts can find exercising their skills enjoyable, becoming absorbed in the task in hand unless it is too easy (Kellogg 1995), such feelings indicating interest. Experts also have a good and well-interconnected knowledge base (Patel and Ramoni, 1997) in contrast to the incomplete, more loosely organised domain-specific knowledge of novices, with difficulties in linking to all their relevant information, including related principles and methods of application (Lesgold, 1988 p202). This should affect their general view of *What is mathematics?* The novices were asked *Did you find that the question interested you?* and *Does the question seem to belong to mathematics?* and to respond on a 3 point scale (no:1, not sure:2, yes:3), and, further, to give reasons for views expressed. In each case, the qualitative responses were categorised revealing a range of responses on that demonstrated: positive or negative *interest* in the question (table 3); positive or negative reasons for the question being *in mathematics* (table 4). Dealing first with the interest descriptors (table 3) for the three analogue pairs, the strongest expressions of interest by novices were in questions 5 and 6, which dealt with queuing situations for airline check-ins and for supermarket checkouts. They also showed ambivalence towards sunflower growth and car speed (questions 11 and 12) and a distinct lack of interest in the clearance problem presented by the bridge and the power cable (questions 15 and 16). In these novice responses the ratios of positive to negative interest descriptors for the three analogue pairs were 2.33, 1.0 and 0.38

respectively (n = 11 or 12). The expert responses indicate a very strong interest in the two pairs questions 5 and 6 and questions 15 and 16. The ambivalence noted in the novices' behaviour on questions 11 and 12 is present amongst the experts, but their high level of interest in questions 15 and 16 is in direct contrast to that of the novices.

descriptors of positive interest	descriptors of negative interest
linked to other models/ interested in finding out the model/ part of modelling	not linked to or part of modelling
personal interest/professional interest	no personal interest/professional interest
realistic or useful/related to real life	not realistic or not useful
easy	too difficult/not clear/confusing
challenging/ a lot to think about	straightforward/ boring
mathematically interesting to me	not in maths/in another subject or topic area/ do not like or not good at this maths area/does not include maths calculations

Table 3 Classification of interest descriptors

positive reasons for *in mathematics*	negative reasons for *in mathematics*
uses reasoning	uses common sense/ reasoning / practical experience/ is about real world/ descriptive
uses numbers/ maths techniques/ maths notation/ equations/ calculations/ variables	does not use numbers etc/ uses very simple maths/ concerned with variables
uses an area of mathematics, or a maths related area e.g. algebra	in another subject or topic area e.g. engineering, mechanics
is in mathematics but no reason given	not in mathematics but no reason given
mathematics is involved in modelling/ mathematics comes at a later stage	no relationship with mathematics or mathematical modelling

Table 4 Classification of positive/negative reasons for *in mathematics*

For the novices, questions 5 and 6, situated early in the modelling cycle, have a much weaker perceived link with mathematics than those later in the cycle (questions 11 and 12 and 15 and 16). This is not surprising given their technical nature. The ratios of positive to negative reasons for *in mathematics* for the three analogue pairs were 1.25, 5.0 and 11.0 respectively (n = 11 or 12). The experts' responses were all positive except for question 11 where there was some uncertainty in relating sunflower growth.

In summary, the qualitative results for novices show that whilst questions 5 and 6 are perceived as interesting they are less likely to be regarded as within mathematics and that a keener interest in problems is accompanied by a tendency to classify them outside mathematics. Conversely, in questions 15 and 16, which are located firmly in mathematics, the level of interest is demonstrably lower. The quantitative scale responses were analysed for all 16 multiple-choice questions, questions 1-6 lending weight to the proposition that:

if it's interesting then it's possibly not mathematics,
and questions 9-16 to the proposition that:

if it's mathematics then it's probably not interesting.

5. Concluding remarks and implications for teaching and research

We have established process descriptors (table 1) of the way in which undergraduate students and research students engage with mathematical modelling and provided examples from different stages of the modelling cycle. In doing so, we have addressed broader questions of interest in and location of modelling within mathematics. Previously, Haines et al (2000,2001) had noted that a stronger focus on the early stages of the modelling cycle: assumptions, clarifying questions, mathematisation is required. Further, it is necessary to emphasise the processes that help students in moving from the real world to the mathematical model and vice versa. Zeitz (1997 p. 60) points to the pedagogical importance of helping students to link real-world phenomena and formalisms in a coherent manner and we provide further insights into why such refocusing is needed. It is clear that the learning experiences of the undergraduate students, thus far, have developed perceptions in students of *what is* or *what is not* mathematics. These views directly oppose *interest* levels impacting upon motivation, confidence and achievement in mathematical modelling. We have reported also some expert behaviours amongst novices, for example in the way that distractors are discarded (§3), but so too there are continuing difficulties, which are to be expected, with the breadth of the knowledge base in novices which does not provide the range of experience and applicability necessary for the successful modeller.

Ericsson et al. (1993, 1994, 1997) maintain that experts become so due to extensive, motivated practice in the appropriate domain. They promote focussed, effortful practice on individualized tasks, chosen by a qualified teacher with attainable learning goals. Edwards and Hamson (1996) provide such tasks, which could be used for mathematical modelling, but perhaps embedded in the mathematics curriculum and prior to tackling major projects on a single application that require extensive and extended mathematical modelling skills. Kellogg (1995) stresses the importance of designing practice activities that focus on proven methods and we know that successful methods include periods of brief instruction followed by a period of motivated and effortful adaptation with immediate feedback on performance (Ericsson, 1996). What is more difficult is proving that such methods work in mathematical modelling and this continues to motivate the construction and testing of a comprehensive mathematical modelling scale.

Note 1

Case study questions 1-12 are presented in Haines et al (2000); questions 5 and 6 can also be found in Haines and Crouch (2001); questions 15 and 16 were inspired by de Bock and Roelens (1993).

References

De Bock D and Roelens M (1993) 'The Journey of the Drilling Rig Yatzy: Today on television, tomorrow as a large-scale modelling exercise' in de Lange J et al (Eds) *Innovation in Maths education by Modelling and Applications*, Chichester: Ellis Horwood, 211-224.

Chi MTH, Feltovich PJ and Glaser R (1981) 'Categorization and representation of physics problems by experts and novices' *Cognitive Science*, 5(2), 121-152.

Edwards D and Hamson M (1996) 'Mathematical Modelling Skills' London: Macmillan.

Ericsson KA (1996) 'The Acquisition of Expert Performance: An Introduction to Some of the Issues' in Ericsson KA (Ed) *The Road to Excellence*, Mahwah NJ: Lawrence Erlbaum Associates, 1-50.

Ericsson KA and Charness N (1994) 'Expert Performance: Its Structure and Acquisition' *AmericanPsychologist*, 49, 725-747.

Ericsson KA and Charness N (1997) 'Cognitive and Developmental Factors in Expert Performance' in Feltovich PJ et al (Eds) *Expertise in Context*, Menlo Park, California. AAAI Press/The MIT Press, 3-42.

Ericsson KA, Krampe, R Th and Tesch-Rmer C (1993) 'The Role of Deliberate Practice in the Acquisition of Expert Performance' *Psychological Review*, 100(3), 363-406.

Galbraith PL and Stillman G (2001) 'Assumptions and context: Pursuing their role in modelling activity' in Matos JF et al (Eds) *Modelling and Mathematics Education ICTMA9: Applications in Science and Technology*, Chichester: Horwood Publishing, 300-310.

Glaser R (1996) 'Changing the Agency for Learning: Acquiring Expert Performance' in Ericsson, K A (Ed) *The Road to Excellence*, Mahwah NJ: Lawrence Erlbaum Associates, 303-312

Glaser R and Chi MTH (1988) 'Overview' in Chi MTH, Glaser R and Farr MJ (Eds) *The Nature of Expertise*, Hillsdale, NJ: Lawrence Erlbaum Associates, xv-xxviii.

Haines CR, Crouch RM and Davis J (2000) 'Mathematical Modelling Skills: A Research Instrument' Technical Report No.55, Department of Mathematics. Hatfield: University of Hertfordshire, 23pp.

Haines CR, Crouch RM and Davis J (2001) 'Understanding Students' Modelling Skills' in Matos JF et al (Eds) *Modelling and Mathematics Education ICTMA9: Applications in Science and Technology*, Chichester: Horwood Publishing, 366-380.

Haines CR and Crouch RM (2001) 'Recognising Constructs within Mathematical Modelling' *Teaching Mathematics and its Applications*, 20, 3, 129-138.

Heyworth RM (1999) 'Procedural and conceptual knowledge of expert and novice students for the solving of a basic problem in chemistry' *International Journal of Science Education*, 24 (2), 195-211.

Kellogg R (1995) 'Cognitive Psychology' London: Sage Publications, 210-211.

Koedinger KR and Anderson JR (1990) 'Abstract Planning and Perceptual Chunks: Elements of Expertise in Geometry' *Cognitive Science*, 14, 511-550.

Lesgold A (1988) 'Problem Solving' in Sternberg RJ and Smith EE (Eds) *The Psychology of Human Thought*, Cambridge: Cambridge University Press, 188-213.

Linacre JM (1990) 'Modelling rating scales' paper presented at the Annual Meeting of the American Educational Research Association, Boston MA., USA, 16-20 April 1990 (ED318 803).

Patel L, Kaufman DR and Magder SA(1996) 'The Acquisition of Medical Expertise in Complex Dynamic Environments' in Ericsson KA (Ed) *The Road to Excellenc*, Mahwah NJ: Lawrence Erlbaum Associates, 127-166.

Patel VL and Ramoni MF (1997) 'Cognitive Models of Directional Influence in Expert Medical Reasoning' in Feltovich PJ et al (Eds) *Expertise in Context*, Menlo Park, California: AAAI Press/The MIT Press, 67-99.

Priest AG and Lindsay RO (1992) 'New light on novice-expert differences in physics problem-solving', *British Journal of Psychology*, 83, 389-405.

Schoenfeld AH (1987) 'What's all the fuss about metacognition?' in Schoenfeld AH (Ed) *Cognitive Science and Mathematics Education*, Hillsdale NJ USA: Lawrence Erlbaum, 189-215

Sternberg RJ (1997) 'Cognitive Conceptions of Expertise' in Feltovich PJ et al (Eds) *Expertise in Context*, Menlo Park, California: AAAI Press/The MIT Press, 149-162.

Zeitz CM (1997) 'Some Concrete Advantages of Abstraction: How Experts' Representations Facilitate Reasoning' in Feltovich PJ et al (Eds) *Expertise in Context*, Menlo Park, California: AAAI Press/The MIT Press, 43-65.

5

Investigating Students' Modelling Skills

Ken Houston
University of Ulster, Northern Ireland
sk.houston@ulster.ac.uk

Neville Neill
University of Ulster, Northern Ireland
nt.neill@ulster.ac.uk

Abstract
The teaching of mathematical modelling is an important part of many third-level curricula. While the methodology of such teaching is well documented, much less work has been done on determining exactly how students perceive modelling situations. This paper reports on continuing attempts to address this area.

1. Introduction

Recently Haines et al (2001) and Haines and Crouch (2001) have presented the results of an investigation into the understanding of students' modelling skills. These authors kindly gave us pre-publication copies of their papers and copies of the test instruments they used.

Haines, Crouch and their co-workers had the objective of developing a simple test instrument (actually a set of two, comparable, test instruments) which could give insights into some aspects of a student's modelling skills. They concentrated on the aspects of "making simplifying assumptions and moving from the real world to a mathematical world", on "formulating a mathematical model", and on "interpreting the mathematical output of a calculation in terms of the original real world problem".

The comparable test instruments each comprised six problems. The problems dealt respectively with the following steps in the modelling process: - selecting assumptions, asking clarifying questions, stating the mathematical problem, selecting variables, formulating the mathematical problem and interpreting possible mathematical solutions. Five possible answers were given for each question, one of which was deemed to be the "correct" answer by the authors and scored 2 marks if selected, and another was deemed to be a "partially correct" answer which scored 1 mark if selected. The authors gave a rationale for their choices of correct answers, partially correct answers and distractors. It is intended that one of the instruments, Test A, could be used as a pre-test to establish a student's initial benchmark, while Test B could be used as a post test, to measure the change in the student's modelling skills. The tests are interchangeable.

Details of the tests may be found in the papers by Haines et al and are not reproduced here.

Haines et al reported at ICTMA 9 the outcomes of a small pilot study using these tests with 39 students and at ICME 9 reported the outcomes of a further investigation, using a composite of both tests, with 42 students. At ICTMA 9 they said "The questionnaire has been shown to be effective in that it is possible to obtain a snapshot of students' skills at key development stages without the students carrying out a complete modelling exercise. The individual questions and the tests as a whole require further and extensive field trials to establish the equivalence between Tests A and B, to test their reliability and to address the effectiveness of the chosen distractors."

The authors have held further field trials, as suggested by Haines and Crouch, to obtain more data on which to attempt a more detailed analysis of this research instrument. Additional work in this area is also reported in Haines, Crouch and Fitzharris (2003).

2. The subjects of the trial

There are two mathematics programmes at Ulster: a two-year Higher National Diploma (HND) course in Mathematical Studies and a four-year BSc (Honours) degree course in Mathematics, Statistics and Computing. The third year of the degree course is spent in industry, usually in a statistical or statistical computing environment.

The entry levels of the two programmes are different with the degree students leaving school with better qualifications. There is some common teaching between the courses, and students who complete the HND successfully can gain entry to the second year of the degree course. The structure of these courses is described in more detail by Challis et al (2002).

The numbers of students in each of the year groups of the courses are given in Table 1 and overall 156 different students were involved. The numbers in BSc2 are a combination of those successful students proceeding from BSc1 and the direct entrants from HND2.

	HND1	HND2	BSc1	BSc2	BSc3	BSc4	Total
1999/00	21	12	18	30	18	19	118
2000/01	11	14	28	21	28	18	120

Table 1 Numbers of students on the courses

3. Course content

Mathematical (including statistical) modelling is a central theme of both courses at Ulster. The second year HND and first year BSc cohorts take a common module in this area. Students work in groups on two coursework tasks. One task requires each group to research a selected mathematical model, using suggested sources, to write "notes" on their work for the benefit of their peers, and to present a seminar to the class. The seminar dates are predetermined and usually fall between weeks 5 and 10 of the 12 week semester. The other task requires each group to carry out a modelling project on a topic selected from a list, each group selecting a different topic. A group then writes a report on their work and presents their findings at a poster session at the end of the semester. Each group selects the submission date of this task and they are advised to select a date which complements the date of their seminar. Students also undertake a comprehension test on a pre-studied modelling paper. Both of these cohorts also take a common module in Operations Research, which introduces them to model formulation and solution via standard techniques, for example linear programming, transportation and assignment models and decision theory. Students work both individually and in a group as part of the assessment schedule and the module incorporates lectures, tutorials and laboratory classes.

During the second year of the BSc course the students take a second module in mathematical modelling, this time in the form of two case studies. No formal lectures are given but seminars on report writing and communication skills help consolidate work done earlier in the course. One of the studies is deterministic while the other is probabilistic. Again the class is divided into groups with the groups changing for the second investigation. Their results are formally written up and each group gives an oral presentation of their findings before their peers and members of staff.

The third year of the BSc course is spent in an industrial or commercial organisation. Students are prepared for this important year by means of a Preparation for Placement module, taken by the Placement Tutor for the course and are visited twice during the year. This placement year is organised by the University in conjunction with its employer partners and the majority of students work in the fields of statistical analysis and/or programming. The assessment for the year includes a substantial contribution from the employer, many of whom now look upon placement as a integral part of their graduate recruitment process. There is a notable increase in both maturity and commitment in the returning students, several of whom will have been offered posts with their placement company upon graduation.

In the final year of the BSc the Statistics module and the individual project module are compulsory. The project allows a student to undertake an in-depth study of a topic, to produce a substantial written report on their work and to discuss their findings in a viva voce examination before a panel of staff members. Students make up the requisite six modules from a wide range of options with their selection reflecting their preferred career path.

4. Modifications and application of the tests

It was the authors' intention to use Test A as a pre-test and Test B as a post-test, over two complete academic years and hence obtain a snapshot of student ability across the whole spectrum. The tests were modified by splitting question 6 into two parts. A new question number 3 was introduced which asked students to sketch the graph of a function which would model a situation, and in question 7, following Haines et al, students were asked to select a function which would model the same situation. This provided an opportunity to compare students' graphical and algebraic knowledge of functions.

All students, including those on placement who were sent copies by post, took Test A during the first week of term in October 1999. All then took Test B in April 2000, near the end of the academic year.

When it came to the second year of investigation it was decided to write a new test, Test C, which would be comparable to the other Tests. Test C is given in the Appendix to this paper. In October 2000 the new entrants to both the HND and BSc took Test A while Test C was applied to the remainder of the cohorts namely HND2, BSc2, BSc3 and BSc4. To see whether there was consistency in the results produced, all students took Test B in April 2001. Clearly this meant that the majority of respondents were taking this test for the second time. Solutions to this test, taken in April 2000, had not been discussed with, nor made available to, the students.

The time allowed for each test was 20 minutes and use of calculators was not permitted. In fact only graphing calculators would have been of use to students and then only in question 7. Those students on placement were asked to apply the same restrictions but obviously it was not possible to monitor if they did so or not. Interviews with students on their return from placement suggest that they did indeed follow the guidelines specified in the letter sent to them.

5. Summary of results

The maximum score for each test was 14 with each question having a correct answer, which scored 2 and a partially correct answer, which scored 1. The other three possibilities scored 0.

As well as summing and averaging the marks, the relationship between questions 3 and 7 was investigated in some depth. In addition to awarding marks for the answers given, a note was kept of the type of graph sketched for question 3 and the equation chosen in question 7. This was to ascertain whether students were consistent in their solutions to these two questions - i.e. did the graph match the equation chosen? Thus a student who incorrectly drew a straight line for question 3 and selected the straight line equation in question 7 was flagged as being consistent as was a student who had chosen the correct solutions to both questions.

Table 2 brings together the mean scores for each cohort in each year.

	1999/2000			2000/2001		
BSc1	Test A Test B 8.13 9.18		% Rise 12.9	Test A Test B 7.39 9.81		% Rise 32.7
BSc2	Test A Test B 9.68 10.3		% Rise 6.4	Test C Test B 8.8 9.39		% Rise 6.7
BSc3	Test A Test B 7.85 9.43		% Rise 20.1	Test C Test B 8.85 10.0		% Rise 13.0
BSc4	Test A Test B 8.94 9.33		% Rise 4.4	Test C Test B 6.31 8.94		% Rise 41.7
HND1	Test A Test B 8.05 8.24		% Rise 2.4	Test A Test B 6.67 7.88		% Rise 18.1
HND2	Test A Test B 7.90 10.8		% Rise 36.7	Test C Test B 6.75 8.54		% Rise 26.5
All cohorts combined	Test A Test B 8.60 9.50		% Rise 10.5	Test A/C Test B 7.60 9.30		% Rise 22.4

Table 2 Summary test results

These results indicate that, generally, students scored better on the post-test, Test B, than on the pre-rest, Test A or Test C. It is disappointing to note that students seem to have regressed between years, especially those returning from placement and entering BSc4

The idea of consistency in students' ability to represent a situation both graphically and analytically was investigated by examining the answers to questions 3 and 7. The results are given in Table 3.

	1999/2000		2000/2001	
BSc1	Test A	Test B	Test A	Test B
Consistent	6	3	7	12
Inconsistent	10	14	21	14
BSc2	Test A	Test B	Test C	Test B
Consistent	13	10	6	16
Inconsistent	12	15	14	2
BSc3	Test A	Test B	Test C	Test B
Consistent	5	7	8	17
Inconsistent	8	7	12	6
BSc4	Test A	Test B	Test C	Test B
Consistent	3	4	5	8
Inconsistent	13	13	11	9
HND1	Test A	Test B	Test A	Test B
Consistent	5	7	1	3
Inconsistent	16	10	11	5
HND2	Test A	Test B	Test C	Test B
Consistent	0	3	3	8
Inconsistent	10	6	9	5
Totals	Test A	Test B	Test A/C	Test B
Consistent	32	34	30	64
Inconsistent	69	65	78	41

Table 3 Consistency between questions 3 and 7

The initial impressions indicated by these results are the obvious difficulties students have when trying to describe physical situations graphically. Even when they could produce a reasonable sketch they were often unable to identify its equation mathematically. A positive feature of Table 3 is the trend for improvement in Test B by succeeding year groups e.g. only 3 students out of 17 in BSc1 in 1999/2000 were consistent while the figure rose to 16 out of 18 in 2000/2001. Similarly the BSc2 cohort in 1999/2000 could only demonstrate 10 consistent members out of 25 whereas this became 17 out of 23 the following year. Overall however the figures are such that they do indeed give cause for concern.

6. Analysis of results by gender

Traditionally the mathematics courses at Ulster have had more female students than male. In 1999/2000 there were 71 females and 43 males included in the tests while the figures for 2000/2001 were 66 and 53 respectively. An analysis by gender showed that the differences between males and females were not statistically significant however. Nevertheless it was something of a surprise to note that the male students had performed better, on average, than their female counterparts in all tests..
In the taught modules the females had normally produced better results, often through a more diligent and committed approach to academic life. Perhaps modelling is one area in which male students can grasp fundamental concepts more clearly than their female peers.

7. Do students' opinions agree with those of the "experts"?

Tables 4, 5 and 6 show the frequencies of response for each of the questions.

	Q1	Q2	Q4	Q5	Q6	Q7
Correct	54	4	90	87	98	49
Partially correct	62	85	3	20	34	34
Third choice	12	26	28	17	7	26
Rest	13	24	20	17	2	29

Table 4   Overall results by question for Test A

	Q1	Q2	Q4	Q5	Q6	Q7
Correct	110	38	163	159	183	72
Partially correct	21	69	19	26	6	68
Third choice	45	64	12	7	10	35
Rest	26	33	9	10	5	33

Table 5   Overall results by question for Test B

	Q1	Q2	Q4	Q5	Q6	Q7
Correct	27	23	24	60	49	16
Partially correct	28	16	21	2	5	15
Third choice	7	25	17	3	13	17
Rest	6	4	6	3	1	22

Table 6   Overall results by question for Test C

It can be seen from the tables above that questions 4, 5 and 6 were consistently well done. These questions dealt, respectively, with the formulation of a precise problem statement, identifying variables, parameters and constants and describing a situation mathematically. The students seemed confident with these types of problem.

In question 1, which concentrates on the assumptions underpinning a mathematical model, most respondents obtained either the correct or the partially correct answer.

The results of question 2, which tests the ability to clarify what is to be accomplished by a model, showed marked divergence from the opinion of the "experts". In Test A only four out of approximately 140 candidates chose the optimum solution and the same disagreement is evident in Tests B and C although to a lesser degree.

Question 7 asks the modeller to reflect upon the mathematical model to be used in a particular situation. All three test results show considerable uncertainty in this area.

8. Conclusions

The modelling tests, carried out over a two-year period with six different cohorts in two different courses, proved both interesting and informative. Perhaps the most noticeable result was the divergence between the "experts" and the students in some questions. In informal discussions the students explained the reasons for their choices and this stimulated healthy debate about the mathematics underlying everyday situations.

There was not, as had been anticipated, an increase in scores as the students progressed through the programmes. Despite having taken several modelling modules and worked in industry for a year, the final year degree students did not perform significantly better than their more junior peers, indeed their attempts at some of the questions were very disappointing.

One encouraging feature was the improvement shown by all cohorts during each of the two years of the project. Perhaps such results were to be expected but, even so, the percentage increases in the scores at the beginning and end of the years were often in double figures.

The area, which has been clearly demonstrated as being weakest, is the ability of undergraduates to express real-life problems graphically. This, combined with their evident lack of knowledge of standard mathematical functions, is something that will be addressed in subsequent years.

References

Challis N, Gretton H, Neill N and Houston K (2002) 'Developing Transferable Skills - Preparation for Employment' in Khan P and Kyle J (Eds) *Effective Learning and Teaching in Mathematics and its Applications*, London: Kogan Page, 79 - 91.

Haines CR, Crouch R and Davis J (2001) 'Understanding Students' Modelling Skills' in Matos JF et al (Eds) *Modelling, Applications and Mathematics Education - Trends and Issues*, Chichester: Horwood Publications, 366 -380. Also published in 2000 as 'Mathematical Modelling Skills: A Research Instrument', Technical report No. 55, Department of Mathematics. Hadfield: University of Hertfordshire, 23 pp.

Haines CR and Crouch R (2001) 'Recognising Constructs within Mathematical Modelling' paper presented at ICME 9, Tokyo, and published in *Teaching Mathematics and its Applications*, 20, 3, 129 - 138.

Haines CR, Crouch R and Fitzharris A (2003) 'Deconstructing Mathematical Modelling: Approaches to Problem Solving' in Ye et al (Eds) *Mathematical Modelling in Education and Culture*, Chichester: Horwood, 41-53.

Appendix   *TEST C (Correct solution in bold, partially correct underlined)*

1. Consider the real world problem (do not try to solve it!):

A pedestrian crossing is being considered for a busy road. Assume that the road is a straight one-way single carriageway.

Which one of the following assumptions do you consider the least important in formulating a simple mathematical model which would determine whether the crossing was needed?

A. The crossing will be controlled by buttons pushed by users

B. The density of traffic is constant

C. The speed of traffic is constant and equal to the speed limit

D. Pedestrians cross at a constant rate

E. Pedestrians will not walk long distances to use it

2. Consider the real world problem (do not try to solve it!):

You wish to reverse your car into a gap in a line of parked cars. The space available is approximately half as long again as the length of your car. Which one of the following variables is most important in carrying out the manoeuvre successfully?

A. The turning radius of the car

B. The distance you drive past the space

C. The prevailing weather conditions

D. Whether or not you can mount the kerb

E. The distance between your car and the parallel parked cars when beginning to reverse

3. An object is dropped from a tall building. Sketch the graph of the distance fallen against the time elapsed

4. Consider the real world problem (do not try to solve it!):

A high street bank has a number of teller-windows at which business may be conducted. Some customers have only one transaction to complete, for example, cashing or lodging a cheque. Other customers have several items of business to transact which may take a long time, such as lodging many bags of coins. Should the bank have a single queue system for serving customers or should it reserve some windows for customers with just a small number of transactions?

In the following unfinished problem statement which one of the five options should be used to complete the statement?

Given that there are six teller windows and given that customers arrive in the bank at regular intervals with a random number of transactions to complete, find by simulation methods the average waiting time for each customer when there is a "single queue - six server" system and compare it with

A. the average waiting time for each customer where separate queues are formed at each window

B. the average waiting time for each customer where separate queues are formed at some windows and a single queue is used for the remaining windows

C. the average waiting time for each customer where a "fast line" queue is formed at one window and a single queue is used for the remaining five windows

D. the average waiting time for each customer where a "fast line" queue is served by some windows and a "slow line" queue is served by the other windows

E. the average waiting time for each customer where queues are formed at two windows and a single queue uses the remaining windows

5. Consider the real world problem (do not try to solve it!):

The University holds regular fire drills to estimate emergency evacuation times. Consider the situation in which students leave a laboratory in single file.

Which one of the following options contain parameters, variables or constants each of which should be included in a mathematical model of the evacuation?

A. Time elapsed after the alarm sounded: Number of students evacuated at time t: Whether the evacuation was in the morning or afternoon.

B. Total number of students to be evacuated: Time elapsed after the alarm sounded: Number of students evacuated at time t.

C. Number of students evacuated at time t: Whether the evacuation was in the morning or afternoon: Width of the laboratory doors.

D. Total time to evacuate all students: Distance between consecutive students leaving: Width of the laboratory doors.

E. Rate at which students leave the laboratory: Initial delay before the first person can leave: Quantity of bags and books carried out.

6. Consider the problem:

New printers are to be purchased for the Computer Services terminal room. The Alpha printer costs £p each and the Beta printer £q each. The Alpha needs r m² of floor space and the Beta s m². Total floor space available is t m². At least b of each type must be bought and the total budget must not exceed £A.

Which one of these options models the situation mathematically ?

A. $x \geq b, y \geq b$ $xr + sy \leq t$ **subject to** $px + qy \leq A$

B. $x > b, y > b$ $xr + sy < t$ subject to $px + qy < A$

C. $x \geq b, y \geq b$ $xs + ry \leq t$ subject to $py + qx \leq A$

D. $x > b, y > b$ $(x+y)(r+s) \leq t$ subject to $(p+q)(r+s) = A$

E. $x \leq b, y \leq b$ $xr + sy \geq t$ subject to $py + qx \geq A$

7. Which one of the following options most closely models the distance fallen by an object released from a tall building (in terms of time t)?

A. $e^{5t} - 1$ B. $(1 - 5t)^2$ C. $5t$ **D. $5t^2$**   E. $\dfrac{1}{1 + e^{5t}}$

6

"How to Model Mathematically" Table and its Applications

Wang Geng
Nanjing University of Economics, Nanjing, China, 210003
w_geng0912@sina.com

Abstract

In this paper, firstly, from the author's own teaching and researching experience, the table "How to model Mathematically" is constructed. Secondly, the author explains the table. Finally, the example, "Walking in the rain" is included to illustrate the use of the table.

1. The "How to Model Mathematically" Table

Mathematics modelling is a process for constructing a mathematical model for a real problem. Here, real problems are of various kinds (dealing with quite wide practice), and mathematics is generalized mathematics and so mathematical modelling is a complicated process. A particular version of the modelling flow chart (see fig.1) is very simple. This chart cannot satisfy people's needs but recall the "How to solve problems" table constructed by the outstanding American mathematician G. Polya in 1944. Can we construct a "How to model mathematically" table in a similarly way?

At present, work like this is not published. The "How to model mathematically" table is presented in Fig 2.

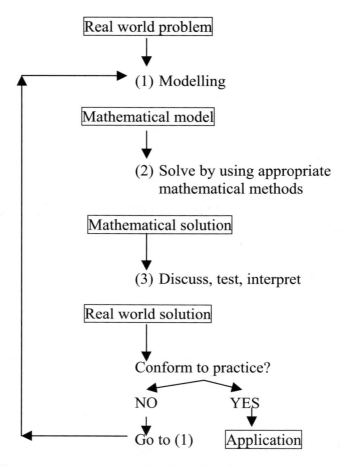

Figure1: Usual mathematical modelling flow chart

"How to Model Mathematically" Table

Stage	Name(Key point)	Objective	Suggestion
Step 1	**Choose Problem** 1. Difficult/Easy 2. Self background such as: Modelling knowledge, Computer ability, Understanding ability, background	**Confirm suitability of** **problem**	Simple simulation Analytical method, Experience method

| Step 2 | **Distinguish problem**
(1)What do we know already?
Data, Relation, Fact, Variety correlation factor and so on
(2)Query problem dealing with knowledge, background, information
(3)Is it credible?
(4)Classify Problem : Certainty/ Randomness/ Fuzzy,
White/ Gray/ Black
Keystone: Explain/ Predict/ Control
(5)Each part related to math.: Content, Arithmetic/ Method
(6)Problematic property: Finite / Infinite/ NP?
(7)Which kind of solution? Data/ Fig./ Relation/ Plan/ Decision and so on
(8)Do we need to use analogy or simulation? | **Complete**
1.Problem statement
2.Problem analysis
3.Assumption
4.Problem character/ Correlation model
5.Mastery corresponding math. knowledge, method class | Firstly, Simplicity;
Secondly, Embed step by step;
Break up the whole into parts;
From elementary to advanced investigate method. |

Step 3	**Work out modeling plan** 1.Search simplest model /Generalize base on given model; Discuss/ Graphic solution. 2.Symbols and units, Compare model and practice problem, list correlation factor, collect data 3.Analyse and confirm necessary assumption, Establish relation and equation 4.Choose suitability math. way design model	**Establish math. model** According to certain "Law", we strike up clear math. problem between variables and parameters	Special method (Need pivot) (Such as: AHP, OR, Neural Networks, Genetic algorithms, and so on) General method

Step 4	**Study problem and obtain solution** (1)Try elementary/ Higher math. methods; (2)Apply appropriate technique /discrete math./ stat. math. method (3)Apply numerical value compute/ Approximately (4)Apply software package/program compute	**To solve the model** Obtain solution and result Data, table, graph, function, formula	Using variety math. methods, Arithmetic, variety math. software (such as: Mathematica, Matlab, Maple, MathCAD, SAS, SPSS, Lindo, QM, and so on)

Step 5	**Translate mathematics solution** (1)Check result and calculation against reason (2)Suggest "Best" solution (3)Consider the whole situation	**Describe** Interpret math. solution	Try putting each data into the model/ on each case/on each condition
Step 6	**Testing, Optimizing, Extending results** 1.Test: Stability analysis (1)Stat. test and Error analysis (2)Compare new model with old one (3)Check practice feasibility 2.Improve: (1)Maximum degradation math. Complexity (2)Conform to practice (3)Consider subordination factor influence for result (4)Consider part important accidental factor influence 3.Generalize: (1)Generally suitable (2)Weaken correlation for its data 4.Evaluate: (1)Veracity (2)Practicability (3)Creativity (4)Compare with standard	**For** (1)Correctness of the model (2)Model relates to experience (3)If dissatisfied, return to step 2. (4)Strengths and weaknesses of the model.	Visual test method Forecast method Random math. AHP. Stat. Synthesize evaluate

Step 7	Report writing and print	The work is completed /proof sheet	Format (Abstract, Problem statement, Assumptions, Symbol, System, Analysis of Problem, Model Design, Model Testing, Strengths, Weaknesses, development of the model references, Appendix)

Fig. 2 "How to do modelling" table

Notes:
(1) AHP—Analytic Hierarchy Process, OR—Operations Research, Lindo—(Linear, INteractive, and Discrete Optimizer) software, MCM—Mathematical Contest in Modeling, SPSS—Statistical Package for Social Science, SAS—Statistical Analysis System, Matlab—Matrix Laboratory, Maple and MathCAD, MathCAI, QM— Mathematical software
(2) Since 1995, about 500 students have used this table; over 90% of students can model simple problems.
(3) Many students' reaction to the Table is similar to their reaction to a teacher who helped them to do modelling.

2. Explanation of the use of the Table and Notes
2.1 Explanation
You can do it according to procedure in the table step by step. At each step, to achieve each purpose refer to main point and point out

method in the table.

The author has applied it effectively for 5 years. Taking advantage of the table, the teacher can guide students effectively to create mathematical models on their own.

2.2 Notes

(1)   Main point, purpose and indicated method may be adjusted on the basis of each objective.

(2)   When we use it, we combine the "How to do modelling" table with the modelling flow chart (Fig.1).

(3)   The table has universality, currency, commonness, It is a guide to mathematical modelling as well as it is a reference to scientific research.

(4)   The table collects together a series of formulization guidelines that are flexible and natural and may be taught easily. This is in accord with the constructivist theory of learning.

(5)   The table is concerned with applying mathematical knowledge.

3. A Suitable Example—"Walking in the rain"

3.1 Background to the problem

It is about to rain; you have to walk a short distance of about 1km between home and college. As you are in a hurry, you do not bother to take a raincoat or umbrella but decide to 'chance it'. Suppose that it now starts to rain heavily and you do not turn back; how wet will you get?

We shall now attempt to model this problem with the help of the table.

3.2 Analysis of Problem (Steps 1 and 2)

Given particular rainfall conditions, can a strategy be devised so that the amount of rain falling on you is minimised? The model will be 'deterministic' since it will depend entirely on the input factors such as the following.

(1) How fast is it raining?

(2) What is the wind direction?

(3) How far is the journey and how fast can you run?

3.3 Assumptions

(1) Speed of rain and rain intensity are constant,

(2) Your speed is constant,

(3) Wind speed is constant,
(4) The human body is considered as a cuboid,
(5) Influence of angle of rainfall is not considered.

3.4 Symbol System
    t — Time during path through rain
    r — Velocity of rain
    θ — Angle of rainfall(due to wind)
    v — your velocity
    Personal dimensions
    h — height
    w — width
    d — Depth
    C — Amount of rain collected on clothes
    I — Rain intensity factor
    D — Distance traveled
    p — Density of rain drop (i.e. Coefficient of rain intensity)
Note: Classifying the Problem :Certainty; problematic solution is Data/ Plan/ Decision

3.5 Model 1:Simple Model (Steps 3 to 6)
Consider simple case i.e. five surfaces to get wet in the rain:
$C = t \times I \times S$
where $t = D/v$ (s), $S = 2w \times h + 2d \times h + w \times d$ (m^2),

We shall need to develop a formula for the amount of rain collected, which is dependent on these factors.

Numerical example: Suppose that the data available are as follows:
    $D = 1000$ m, $h = 1.5$ m, $w = 0.5$ m, $d = 0.2$ m; hence $S = 2.2$ m^2
Also take $I = 2$ cmh^{-1} = $(2 \times 0.01)/3600$ms^{-1}, $v = 6$ ms^{-1}; hence $t = 167$ s
    $C = 2.041$ litre
This is like having about two bottles of wine poured over you! This does not agree with real experience. So we construct a second model.

3.6 Model 2:Advanced Model (Return to Step 2)
Modify assumption (5): Consider Influence of angle of rainfall.
A diagram helps in explaining the situation to be modelled and this is shown in Fig.3, that the rain speed is constant throughout and also that you

move through the rain at a constant speed.

Because the part to be soaked is the top (C_1) and the front (C_2):

$$C = C_1 + C_2 = \frac{pwD}{v}[dr\sin\theta + h(r\cos\theta + v)] \qquad (1)$$

From the earlier data, we have $r = 4$ ms^{-1}, $I = 2$ cmh^{-1} and $p = 1.39 \times 10^{-6}$

Then $C = \frac{6.95\times10^{-4}}{v}(0.8\sin\theta + 6\cos\theta + 1.5v)$ \qquad (2)

Thus, the problem turn into "given θ, what v should be chosen so that C is minimized?"

Step 5

Equations (1) and (2) are now examined. First, note that, if the rain intensity I is zero, then $C = 0$ which means that you stay dry. Second, the value of θ will determine whether the rain is facing you or blowing in from behind. We shall evaluate equation (2) to show what happens in particular cases.

When $\theta = 90°$, i.e. the rain is falling straight down. then

$$C = 6.95 \times 10^{-4}(1.5 + 0.8/v) \quad (\text{m}^3)$$

This expression is smallest for the largest possible value of v, i.e. $v = 6$. Then

$$C = 11.3 \times 10^{-4}(\text{m}^3) \approx 1.131 \quad \text{litre}$$

When $\theta = 60°$, i.e. the rain is driving in towards you. Then

$$C = 6.95 \times 10^{-4}[1.5 + (0.4\sqrt{3} + 3)/v] \quad (\text{m}^3)$$

Again this is at its smallest when $v = 6$, then

$$C = 14.7 \times 10^{-4} = 1.47 \quad \text{litre}$$

When $90° < \theta < 180°$, i.e. it rains on your back, let $\theta = 90° + \alpha$ then

$$C = 6.95 \times 10^{-4}[1.5 + (0.8\cos\alpha - 6\sin\alpha)/v]$$

This expression can become negative for α sufficiently large, which means that the model must be examined more carefully since it is not possible for C to be negative! It is best to return to equation (1) to analyse the situation. There are two cases to consider according to how fast you move through the rain.

Case 1. $v \le r\sin\alpha$

$C = C_{\text{top}} + C_{\text{behind}} = (D/v)wd[pr\sin(90° + \alpha)] + (D/v)wh[p(r\cos(90° - \alpha) - v)]$

$= pwD[dr\cos\alpha + h(r\sin\alpha - v)]/v$

then:

$$C = 6.95 \times 10^{-4}[(0.8\cos\alpha + 6\sin\alpha)/v - 1.5]$$

If $v = r \sin \alpha = 4 \sin \alpha$,
$$C = 6.95 \times 10^{-4} (0.8 \cos \alpha)/(4 \sin \alpha)$$
if $\theta = 120°$ i.e. $\alpha = 30°$, then $v = 4\sin 30° = 2$ (ms⁻¹),
then
$$C = 6.95 \times 10^{-4} (0.8\sqrt{3}/2)/2 \text{ m}^3 = 0.24 \text{ litre.}$$

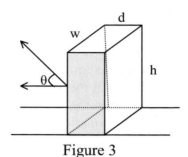

Figure 3

Case 2: $v > r \sin \alpha$
$$C = C_{top} + C_{front} = (D/v)wdpr\sin(90° + \alpha) + (D/v)whp[v - r\cos(90° - \alpha)]$$
$$= pwD[rd\cos \alpha + h(v - r\sin \alpha)]/v$$
When $v = 6$ ms⁻¹ and $\alpha = 30°$, then $C = \times 10^{-4}(0.4\sqrt{3} + 6)/6 \text{ m}^3 = 0.77$ litres

Step 6 Compare with reality
The results seem sensible and agree with what we might expect. In what way has this second detailed model improved on the earlier? Here we have allowed for wind direction and investigated the various cases more thoroughly. All the results for rain collection are less than the value of 2 litres obtained for the initial model. Also the orders of magnitude are what we might expect. It is difficult to validate the numerical results of the model, but the idea of 'moving with the rain' can be tried out in practice, assuming that you do not mind getting wet. The overall conclusions from the model are as follows.
1. If the rain is driving towards you, then the strategy should be simply to run as fast as possible.
2. If the rain is being blown from behind, then you should keep pace with the rain, which means moving with a speed equal to the wind speed.

Note in passing here that the conclusions are given in simple everyday terms that are easily understood. It is no good telling a non-mathematician to run at $r \sin \alpha$ (ms⁻¹).

Step 7 Report writing and print
Omitted from this paper.

References

Li ShangZhi.(1996) 'Tutorial of Mathematics Modelling Contest', Nanking: JiangSun education publishing house.

Edwards D and Hamson M (1989) 'Guide to Mathematical Modelling', London: Macmillan Education Ltd

Polya G (1957) 'How to solve it', New York: Doubleday Anchor Books.

Wang Geng (2000) 'Practicable Computer Mathematical Modelling', HeFei: Anhui university publishing house.

Section B

Mathematical Modelling Competitions

7

New Applications of the Mathematics A-lympiad

Dédé de Haan
Freudenthal Institute, Utrecht University, The Netherlands
D.deHaan@fi.uu.nl

Abstract

The Mathematics A-lympiad is an international competition for teams of 4 students (upper secondary education), organized by the Freudenthal Institute of Utrecht University in the Netherlands. The teams work on a very open ended problem situation in which mathematical problem solving and higher order thinking skills must be used to solve a real world problem. The result of the assignment is a written report.

We can identify a small number of schools in the Netherlands that repeatedly enter the final round. What is it that makes a school successful? To get more insight in these 'successfactors', research has been started. Factors like schoolculture, teachingstrategies, a well thought-out long-range plan in developing skills seem to be important. How to implement these factors in an 'average' school to make it an 'outstanding' school (in scoring in the A-lympiad) is of course the most interesting part of this research.

Different tendencies give rise to interest in these factors and their implementation.

There is the recent curriculum-change in the Netherlands, which forces Dutch math-teachers to assess 'higher order skills' of their students in mathematics. Results of the findings on 'succesfactors' in winning the A-lympiad are used for teacher-training.

Another tendency is that all over the world there is considerable interest in evaluating the attempts to operationalise 'higher order skills'.

The limitations and undesired effects of traditional testing are acknowledged and recognised. Many people are in search of suitable tests/tasks that fit within the bounds of 'fixed time' and 'paper and pencil', but which also attempt to assess the process goals and higher order skills.

1. Mathematics A and the A-lympiad

The Mathematics A-lympiad is an international competition for teams of 4 students (upper secondary education), organized by the Freudenthal Institute of Utrecht University in the Netherlands. The teams work on a very open-ended problem situation in which mathematical problem solving and higher order thinking skills must be used to solve a real world problem. The result of the assignment is a written report.

The competition has two rounds: the preliminary round with about 1000 teams of students competing a day long at their own schools; and an international final in which 16 teams compete during a whole weekend on a different assignment.

1.1 Mathematics A

The Mathematics A-lympiad owes its existence to the subject Mathematics A.

Mathematics A is intended for students who will have little further education in mathematics in their academic studies, but who must be able to use mathematics as an instrument to a certain extent. The emphasis lies on applications of mathematics and mathematical models (more than on 'pure' and abstract mathematics), and also more on the processes to come to an answer than on the answer itself.

However, the 'process'-character of math A didn't get justice in the national final examinations at the end of VWO (grade 12 of pre-university level highschool). And that is a pity, because it may be the case that students do not have an adequate command of the 'lower' calculation skills, but are indeed good at solving problems, at critically considering models, at mathematisation and logical reasoning.

If we presume this assumption is correct (De Lange, 1987), it is good to design a 'task' for mathematics A that attempts to encompass the original objectives of Mathematics A.

1.2 Mathematics A-lympiad

The Mathematics A-lympiad was born!

De Lange (1987) suggested already that designing this kind of open-ended problems is not easy. That's why a committee to design A-lympiad problems, was formed. The committee consisted of people who had won their spurs in the area of Mathematics A: most of them had been involved in developing the curriculum for Mathematics A.

It was decided to offer the problems in a contest-form and it was also decided that every problem was to be handled by a team of students (3 or 4), and that their hand-in work had to be a paper containing all aspects from defining the problem, via strategy definition, solution, argumentation, to presenting the solution.

The contest was a fact: in March 1990, 14 teams of 'lead' schools were invited to come to a conference-center in the woods. They worked for 2 days on an open problem. This first time was a great success! So the next schoolyear ('90-'91), a 'preliminary round' was introduced, to be held in november: the problem was sent to the schools that were interested in the contest, and they arranged that teams worked on the assignment during the day. The best two papers were to send in to the Freudenthal Institute, to compete for the final round. The final round took place in March, and the best 12 teams were invited to participate there.

Gradually, more and more schools, with more and more teams, participated.

The number of participating schools in the Netherlands seems to be stable the last couple of years: about 110 to 120 schools of pre-university level participate. The number of participating teams seems to be decreasing a little.

Since 1995, Denmark also participates in the Mathematics A-lympiad, and two years later the Dutch Antilles in the Caribbean also entered the competition.

Other countries, like for example Germany and Portugal, are interested in competing as well.

2. The characteristics of the contest

Why is this kind of contest so popular?

If we take a look at other parts of the world with similar contests, we see the same phenomenon: for example the HiMCM (Highschool Mathematics Competition in Modeling) organized by COMAP in the USA has a lot of competitors; in China there is a Mathematical Contest in Modeling for High school students called "Competition in Applying

Mathematical Knowledge for High School Students (CAMK)" since 1991 in Shanghai and Beijing, and now in more than 10 provinces.

Apparently, it is not a separate phenomenon; teachers seem to be interested in this kind of contest all over the world. Let us consider some of the characteristics of the A-lympiad:

1) the problem should improve learning
2) the problem should allow candidates to show what they know (positive testing)
3) the problem should operationalize the goals of the Mathematics A curriculum
4) the quality of the student work is not in the first place measured by the accessibility of objective scoring
5) the problem is delivered in a contest-form, to promote participation and for teachers to build up practice in assessing student work
6) group work should promote mathematical understanding and stimulate level raising

The first five are copied from De Lange (1987), which is self-evident, because the Mathematics A-lympiad is a direct consequence of his recommendation to let experts develop alternative tasks.

2.1 Ranking instead of grading

The fourth characteristic implies a difficult job for teachers to assess the student work. Objective scoring is not possible; however, scoring by different graders leads to almost identical rankings. That is why teachers are not asked to grade the student work, but to rank them, from best to worst. The committee provides a number of general and problem specific criteria teachers can use.

In "10 years Math A-lympiad – the real world mathematics team competition" (De Haan and Wijers, 2000), a complete chapter is dedicated to the assessment of answer papers.

2.2 Contest

The fifth characteristic (presenting the problem in contest setting) is first of all meant to create a situation to fit this kind of activity into the school practice (all schools and teams on the same day, within restricted time). Other positive consequences of this setting are that: schools are challenged to compete; it is well-organised: one problem a year; you can choose as a teacher to grade the student papers, or not to grade; the contest can be used as public relations for the school (not only when a school reaches the final round, but also within the school, to honour the team whose answer paper will be sent to the Freudenthal Institute to compete for the final round); it stimulates students, colleagues and parents; because of the way the 'scoring' is set up, all competing teachers will also see answer papers from other students, from other competing schools, which can help in building up experience in assessing answer papers. A more recent consequence of the contest setting is the origin of a network of schools, (every school that enters the competition), with a yearly network-meeting, which is an ideal setting to exchange information and share experiences.

2.3 Group work

The sixth principle mentioned in the list of A-lympiad-characteristics is not emphasized greatly by De Lange, but is an important factor in the A-lympiad.

The focus on the outcomes of group-work can be found in several reports.

What is very interesting is the process model presented by Dekker and Elshout-Mohr, that is meant to show how mathematical level raising for individual learners can be realised by allowing students to work in small groups on mathematical problems (Dekker and Elshout-Mohr, 1998).

3. The process model in group work

The basis of the problems used in the process model presented by Dekker and Elshout-Mohr is that they should be realistic or meaningful to the students, complex (in the sense that several skills are needed to solve the problem) and constructive (producing a graph, table, drawing, story) in order that differences between students become visible and subject to discussion. The problems should also aim at 'level raising', which means that they will develop new mathematical concepts because the old ones are not sufficient anymore.

The process model used by Dekker and Elshout-Mohr is constructed around four key activities:

- To show one's work
- To explain one's work
- To justify one's work
- To reconstruct one's work

When level raising occurs, this will be revealed in the reconstruction of one's work; if the result of reconstruction is shown or told again, the cycle is completed.

The process mentioned above can take place with a student who works alone, but it will take a lot of self-regulation; it's more natural to let the process happen in communication with other students.

3.1 An example

If we look at the way the A-lympiad-problems are designed, we can recognize the principles.

We will look at an example to show the aspects mentioned above, and their realization in practice.

Scratch cards

Scratch cards are popular. They are appealing: just by scratching the right box, you scratch yourself into a millionaire. By choosing what you scratch you can believe you are influencing your own luck.

There are different scratch systems. In spring the Dutch Railways had a scratch card campaign. When you bought a cup of coffee you were given a scratch card with three squares.

You had to scratch two. If you had two flags with an S on them then you won a free return trip with Stena-Line from Hook of Holland to Harwich. If you scratched two cups, then you got a free cup of coffee. If all three had been scratched the card was void.

With other scratch systems you can win more money with each square that you open, until you scratch a square with a cross. Then you lose it all. This of course plays on the greed of the person with the card.

This assignment is about this last system.

Scratch it yourself!

The DIY store **DHZ** will shortly have its 25th birthday. To celebrate this jubilee and to pull in extra customers, the owner decided on the **Scratch it yourself!** scratch card campaign.

For every f 100 (≈USD 40) spent the customer was given a scratch card with 10 squares that could be scratched. By spending f190 you would thus get one card. If you spent f 205 you would get two.

Every card contained five squares with the **DHZ** logo. These were distributed randomly over the 10 squares.

The following rules of play were printed on the card:

```
Rules
With one open square showing the DHZ logo you win ƒ7.50.
With two open squares showing the DHZ logo you win ƒ15.00.
With every subsequent square showing DHZ the amount you
have already won, doubles.
If you scratch a cross, you lose everything.
```

The owner felt that this campaign was going to cost him dear.

One week before starting he did a trial among his staff. He had his employees scratch open 200 cards just as if they were customers. He did not actually pay out the prize money of course.

He got the following results:

52 of the 200 cards had won ƒ7.50. Fourteen people had opened two logos and won ƒ 15. Four cards had won ƒ30, one had ƒ60 and one had ƒ120. He was shocked by the total prize money for these 200 cards.

Task 1

Analyse this campaign. Take the following points into consideration:

- o The campaign speculates on the greed and gambling lust of the customer. How is that reflected in the outcome of the trial?
- o After the trial a number of employees said that they would expect a lot of hassle and moaning from customers. A customer who first opened up a cross and then three logos would then complain that he was entitled to a prize of f 30. Where do you think this fear of the staff comes from?

The owner wanted a different campaign. It had to be less expensive. Nevertheless, the customer had to be able to win back what he spent. The campaign had to be of interest and draw in customers. In addition it should not generate any hassle. He wanted to distribute 10 000 cards. The entire campaign was not to cost him more than 2% of the turnover involved.

Task 2

Devise a different (better, nicer, more sensational) scratch card campaign for the owner. The campaign must satisfy the conditions of the owner:

1. Main prize at least $f100$.
2. Total prizes not more than 2% of turnover.

The owner was open to alternatives. You can thus set up a completely different system.

Show that your design satisfies the requirements of the owner.

Also specify why you think your system is better.

The problem is simply stated and complex to stimulate students to show what they know and to stimulate level raising by group process. This can be done for instance by first working individually and then sharing ideas and explaining to each other. The way the group process takes place will be of influence on the quality of the work.

The problem is coherent and specified: there was an experiment with a certain scratch card campaign; the owner of the shop wants a different scratch card campaign; the reason why is explained; the conditions are clear. This stimulates students to reflect on their work and their assumptions.

The justification of the chosen strategy or solution, determines if the answer paper is bad, good, or better; not the chosen strategy or solution itself.

The subject of the problem and the group process can promote the development of mathematical concepts; in this problem, the concept of probability can be developed without knowing about it before. This is what happened in some groups.

4. Success factors

If different solutions and strategies are possible, there will be different results; the quality of the different solutions will also be different, but what are the conditions to be fullfilled to have an 'excellent' answer paper?

This question came up when, after a couple of years, the same 10 schools appeared in the final round almost every year, (almost) each time with another team. Is there a kind of math teaching and classroom environment that somehow leads to superior competition skills?

Two short research-projects (De Haan and Wijers, 2000; Harskamp, De Haan and Van Streun, 2000) showed that there are indeed factors that can make your team successful in winning/reaching the final round of the Mathematics A-lympiad, or, in other words: to make them successful in using their problem-solving skills and in working as a team: schoolculture and knowing how to teach problemsolving skills.

School culture
  ➢ It may sound trivial but is very important: to develop problem-solving skills of your students, start at an early age, beginning of high school (grade 7 in the Netherlands), with small, more structured and 'closed' investigations; during their high school career, the problems can become bigger and more open, so in their final year (grade 12) students are able to formulate their own investigation, pose the problem etc.
  ➢ It also helps if 'developing (problem-solving) skills' not only takes place in the subject of mathematics, but also in other subjects.
  ➢ It really helps when the mathematics department is enthusiastic, stimulates students to participate, and maybe also stimulates to participate in other contests (in mathematics or other subjects).

Knowing how to teach problem solving skills
  ➢ To get the most out of the team-work, students should learn how to work in groups. There are the groups that have one 'smart guy/girl', one who can type very fast at the computer, and two who take care of the tea and coffee (to give a somewhat extreme example). This kind of group-interaction is not what is meant by Dekker and Elshout-Mohr; they assume participants of the group to first work individually and then to share and talk about their different solutions. That group-process is needed to raise the level of the end-product; the schools that enter the final round of the Mathematics A-lympiad more than once, always send teams of which the members first read through the problem by themselves and then share the ideas they have on possible solutions. So they

raise the chance that they might undergo the process model described by Dekker and Elshout-Mohr, and indeed raise the level of their end-result.

> As already mentioned under 'school culture': start early with good problems! As a teacher, you should be aware of what good problems are and how to handle them in classroom

> In assessing end-products, the emphasis must be on the assessing of the problem-solving skills

5. Changes

Recent changes in the highest grades of Dutch curriculum have caused more interest in these 'success factors', and these curriculum-changes (which are implemented in 1999) have also consequences for the contest itself.

The main curriculum changes are that every student has mathematics as a subject, whereas before students could drop the subject in grade 10-12. The new curriculum requires complex student projects in different subjects, which are part of formal assessment now, so there are many alternative tasks in textbooks now.

So one could say: the A-lympiad has reached its goal and has made itself superfluous.

But this conclusion is drawn too fast.

5.1 Pros and cons of the contest in the new curriculum

There are some factors militating against the contest:

In the new curriculum, students are getting used to these kind of larger tasks, in every subject, and so it is 'normal' to them, and maybe even boring. A lot of teachers are not interested in the competition: it is not exceptional anymore, everybody has to assess these kind of problems; the fact that it is a contest is even very inconvenient: you don't know what the subject will be! But they still want to use the A-lympiad-problems, because in general these are 'good' problems (considering the principles mentioned earlier), and the problems can be done within a day.

Answer papers of the preliminary round are graded by most teachers. Last year only half of the schools that received the problem by applying for the contest, sent in the best answer paper to compete for the final round.

Teenage culture now requires part-time jobs and time to spend the income from the jobs. School has become a lesser priority. We notice that the last couple of years, the interest in entering the final round is decreasing.

On the other hand:

Just right now, the experience of the A-lympiad developers and teachers must be published, so that everybody who wants to know more about practising and assessing higher order thinking skills, can get access to this information.

That is why we keep on organizing the contest: to offer good complex tasks, in our opinion, to teachers; and to keep the network alive, to pass on the knowledge on 'success factors'.

5.2 Implementing success factors in in-service teacher-training

From October 1999 to May 2000, teachers were supervised by researchers to find an efficient way to develop problem-solving skills of their students who were already in grade 10 and had never done this kind of work before (Harskamp, De Haan and Van Streun, 2000). From the field of teachers, there was a demand for this kind of investigation and training, because of the latest changes in the Dutch school curriculum.

A course was developed for teachers in two different schools. In the course, teachers had to design open problems (with the use of explanation and examples), which they had to use and supervise in class. The researchers developed the course and observed in classroom. Three meetings were held with the complete mathematics-faculty of each school.

The outcomes of the research were:

- Communication between the math-faculty is necessary. In one of the schools there was a teacher with a lot of experience in assessing and practicing problem solving; his colleagues learned about this only at the meetings in the framework of this research.
- To develop problem-solving skills of your students, starting from zero, it's best to design the problem in such a way that there is a demand for describing the process of working; preferably in an natural way embedded in the text. There are different tools to use to achieve this; for example, let students play the 'role' of reporter; define (in a creative way) what the product should be; put it in some kind of context. As soon as teachers and students have more experience with this kind of problems, students will know better what is expected from them.

- To supervise the students as effectively as possible, as a teacher you should know beforehand what you want to see as the end-product, and what should at least have been done by the students to achieve that. Only then, you can react in the right way when you see the students working. Ideally, as a teacher you should really know the assignment, and preferably a lot of possible ways of solving it! So when you give the assignment, it's good to look at it from the student point-of-view.

    The more open your assignment is, the more you make an appeal to the problem-solving skills of the students, but also: the greater the chance that it goes wrong completely.

This research confirms the earlier research on success factors. Teachers clearly need to be trained to teach and facilitate problem solving both in pre-service and in-service education.

References

Dekker R and Elshout-Mohr M (1998) 'A process model for interaction and mathematical level raising' *Educational Studies in Mathematics* 36, 303-314.

Haan DL de and M Wijers (2000) '10 Years Math A-lympiad- the real world mathematics team competition' Utrecht, The Netherlands: Freudenthal Institute.

Harskamp E, Haan DL de and van Streun A (2000) 'Praktijkbrochure Praktische opdrachten wiskunde' Groningen/Utrecht, The Netherlands: GION.

Lange J de (1987) 'Mathematics, insight and meaning' Utrecht, The Netherlands: OW and OC

World Wide Web: http://www.fi.uu.nl/Alympiade

8

"Mathematics Contest in Modelling" Problems from Practice

Shang Shouting, Shang Wei and Zheng Tong
Harbin Institute of Technology, China
Shang Shouting <sst@hope.hit.edu.cn>

Abstract

Eight Problems have been designed in the Institute's annual Mathematics Contest of Modelling. These problems are mostly from the designer's own experience or the scientific projects they currently study. Broad thinking space is available for students to exert their creativity and imagination. One of the problems in the 2000 contest is shown in this paper. The solutions and the students' papers are introduced as well.

1. Introduction

COMAP(http://www.comap.com), a Consortium for Mathematics and its Applications, is an award-winning non-profit organization whose mission is to improve mathematics education for students of all ages. COMAP has organized an international mathematics contest of modelling. The participants are undergraduate students from all over the world. International participants get the problem from the internet and submit their papers both via internet and via air mail every year. There are 522 teams from 11 countries in the 2002 U.S.A MCM, among these are 281 teams from U.S.A and 216 teams from China.

In the training for such contests as well as in mathematics modelling and practicing classes, it is necessary to put forward some actual problems for students to exert their own. These problems should be solvable using their existing mathematical knowledge and skills so that they could build mathematical models and find solutions. The Chinese MCM and the USA MCM (2002) are sources of good materials. However, there are published solution papers to these problems. Thus, such problems from past contests cannot be used to develop students' innovation, but can only be used to provide some examples. It is not easy to construct problems without direct paper reference. Moreover, these problems should also be suitable for college student to solve in as little as three days. In the past several years, we have considered and constructed eight problems using our current research and experience. This provides adequate material for the exams of the mathematics experiment and the Harbin Institute's MCM.

1998	A. exploitation plan for oil field output maintenance
	B. analysis and processing of the students' scores
1999	A. solar energy receiver's orientation
	B. problem of cracking passwords
2000	A. racing cars' swerve skills
	B. measurement of coal heap
2001	A. processing image's data
	B. mercury pollution in the Songhua River

Problem 2000 B and the solution are given as follows:

2. Problem: Measurement of large coal heap

Since power plants, thermoelectric stations and central heating systems have a great demand for coal, they always need to estimate the current reserve. Due to huge size of the coal heaps, it obviously could not be weighed using a scuttle or balance. And errors will occur inevitably, if the reserve is estimated only by experience. Your task is to design an easy and executable method to measure and calculate the volume of a large, reserve coal heap. Satisfactory precision should be considered in your solution.

1. Give one or two easy and fast measurement techniques, and the corresponding data processing method. Sensitivity should be analyzed.
2. Based on the obtained data, build a mathematical model to simulate

the coal heap and design the method for calculating the reserve volume, together with a computer program. Errors should be estimated also.
3. Construct a set of data relatively close to the real case, prove your model and make sensitivity and error analysis.

3. Background of the problem
The chief engineer of a power station originally formulated this problem. The station has storage for 300,000 tons of coal. In order to estimate accurately the capacity of the reserve field periodically, it calls for an executable, easy-to-do, and inexpensive method of measuring and calculating. Though there is no need to reach a very high accuracy in estimation, an acceptable error range should be specified. At the time, to capture the data, photographs and videos were taken. Both an image identification technique and a simple measuring instrument can be adopted to obtain the data separately. The former is more difficult in technique, while the latter requires consideration of the relation between workload and accuracy

4. Key to the problem
As a problem designed for college students' mathematics modelling contests, solving the coal heap measurement problem has the following key points:
4.1 Basic assumptions
1. The ground is one of the reference coordinate plan: $Z=0$
2. Surface of the coal heap is a continuous smoothed surface: $Z=Z(X, Y)$
3. Any tangent plane of the coal heap makes an angle $\alpha \leq 55°$ with the ground (Jiang, 2000)

4.2 Capture the data
1.   Direct measurement
Choose a planar meshwork (x_i, y_j), use theodolite or laser to measure the height of the intersections of the meshwork: $Z_{ij}=Z(x_i, y_j)$, give the relevant expressions. The number of measured points relates to the accuracy of the measurement.
2.   Photographic picture analysis method
Make marks on given reference coordinates in the surface of the coal heap, and take photos at different orientations. According to the

relative positions in the photo, decide the vertical coordinate values on the marks.

3.   Some other feasible method

4.3 Data density increasing

Since the volume calculation is simplified by summing up all the products of the unit area and the height of the sample point, the accuracy of the unit area affects the final accuracy greatly. Increasing the data density to get smaller and more precise unit area can obtain more accurate result. There are two ways to increase data density. One is to increase the number of testing points. This method is indeed costly and impracticable according to the given background information. The other method is interpolating the smoothed surface to get some simulative points. This is an important means to reduce the measurement work and enhances the accuracy at the same time (Xiao, 1997)

Let XX, YY, ZZ be density increasing, interp1 and interp2 are the two dimensional and three dimensional interpolated functions provided in the software of Matlab, the 'bicubic' and 'spline' is the specific algorithm used in the function:

a. $ZZ(XX,YY) = \text{interp2}(X, Y, Z, XX, YY, \text{'bicubic'})$,or
b. $ZZ(XX, y_j) = \text{interp1}(X, Z(X, y_j), XX, \text{'spline'})$
    for $j = 1,2,...N$;
    $ZZ(xx_i, YY) = \text{interp1}(Y, Z(xx_i, Y), YY, \text{'spline'})$
    for $i = 1,2,...M$;
    $ZZ = ZZ(XX,YY)$.

4.4 Calculating the volume of the coal heap

Carry out a numeric integration using interpolated data to attain the volume of the coal heap. The trapezoidal method or other two-dimensional integral method can be adopted.

(a)

$$V = \int_c^a \int_a^b Z(x, y)dxdy$$

$$\approx \frac{h_x h_y}{4} \sum_{j=0}^{m-1} \sum_{i=0}^{n-1} [ZZ(x_i y_j) + ZZ(x_i y_{j+1}) + ZZ(x_{i+1} y_j) + ZZ(x_{i+1} y_{j+1})]$$

(b) $S_i = \dfrac{h}{3}(zz_{i,0} + zz_{i,2m} + 4\sum_{k=1}^{m-1} zz_{i,2k+1} + 2\sum_{k=1}^{m-1} zz_{i,2k})$

$V \approx \dfrac{h}{3}(S_0 + S_{2m} + 4\sum_{k=1}^{m-1} S_{2k+1} + 2\sum_{k=1}^{m-1} S_{2k})$

4.5 Error analysis

Based on the adoptive interpolation and integral method, decide the accurate of your calculation. Condition as the basic assumption 3 can be appended.

4.6 Sensitivity analysis

According to the measurement strategy, analyze how the measurement error affects the calculation result.

5 Summary of the students' response

5.1 Students' opinion of the contest

Most students seemed enthusiastic when they got the problem. They discussed hotly within their team and one of them even went to the nearby boiler station to survey the coal stack there. The paper from this team eventually proved to be the best one. Many others showed a lack of practice, creativity and imagination. In spite of this, all of the participants commented that their problem solving ability improved greatly along with their teamwork skills. They introduced the contest to their classmates as a useful experience in undergraduate years.

5.2 Summary of the result papers

5.2.1 Means of measurement

Students adopted many measuring means, and some answers to questions in the test paper are very detailed. The students not only wrote about the models of measurement and calculation, but also offered several kinds of instruments' performance, precision and even market price. Some students went to several nearby coal yards to investigate, and proposed simple and convenient measurement methods like hanging a plumb line.

5.2.2 Data processing

Most students adopted the method of interpolation, constructed the simulation data, used the ready-made software, and made a three-dimensional simulation dump picture. Some solutions using random data did not get a good effect and did not even conform to the meaning of the question.

5.2.3 Numerical integration

The majority adopted the trapezoidal method. Some others use the rectangular method, which cause a considerable loss in calculation precision.

5.2.4 Precision and sensitivity analysis

Only a few students made a relatively detailed analysis, and many answers only gave the error percentage. This part of the solution is one of the main signs for distinguishing the quality of the paper. Though some students have studied a course in numerical analysis, they still do not take it in account. This proves that students are deficient in understanding and using the error analysis, which should attract the teachers' attention in the future.

5.3 General appraise and analysis

According to the students' reflection, the problem has the proper degree of difficulty, and is close to reality. The knowledge required does not go beyond university science mathematics courses. It is effective to improve the students' ability to practice, innovate, model, calculate and analyse. We used it as the homework for modelling or experimental courses many times after the contests, and each time we got new arguments and methods from the students.

On the other hand, however, many students cannot integrate their known knowledge with the given practical problem, especially those new to MCM problem solving. This is caused by the routine mathematics teaching method, which has terrible ignorance of applying mathematical method into real practice. The evidence is that the more a student works for MCM papers, the more creative and scientific his paper will be. We are delighted that these problems are good material to raise students' overall qualities.

References

Jiang Qiyuan (2000) 'The Piling up of the Gangue', 99 the Cup of Chuangwei Mathematics Modelling Contest (The group of junior colleges), the practice and knowing of mathematics, 1, 3-4.

U.S.A. Mathematical Contest in Modelling (MCM) (2002) http://www.comap.com/undergraduate/contest

Xiao Shu Tie (1997) 'Mathematics Experiment', Beijing: High Education Press, 7.

Section C

Using Technology in the Teaching of Modelling

9

Modelling and Spreadsheet Calculation

Mike Keune
Ecumenical Cathedral School of Magdeburg, Germany
M.Keune@gmx.de

Herbert Henning
Department of Mathematics, Otto-von-Guericke University of Magdeburg,
Germany
Herbert.Henning@gmx.de

Abstract

On the basis of observations in the present German educational system and the requirement to include modelling activities as a part of the general aim in mathematical education the advantages of spreadsheets will be demonstrated with the help of three examples.

To start with the assumption that, the earlier one begins with the concept of modelling the better one acquires these abilities in the course of the school time and the better one recognises mathematics as a part of our world, modelling activities in lower secondary schools will be shown. In the modelling process spreadsheets could be used with models which are based on systematic testing, iteration and recursion, simulation, parameter variation, visualisation and data-handling in an elementary fashion.

1 Introduction

To start with modelling activities in lower secondary schools (or even in primary schools) means to start with uncomplicated but exemplary real world problems and increase the complexity in lower and upper secondary schools, until final examinations. This is to achieve one of the most important goals of mathematical education, namely the acquisition of the ability to handle oncoming problems from different parts of life using mathematical methods.

In the last three decades there has been a lot of scientific progress, suggestions and proposals in the world of mathematical modelling in the educational field. The ICTMA-conference series is one good example of this progress, but the every day reality in school is different (probably not only in Germany)! As Gabriele Kaiser from Germany at the ICTMA 9 announced "[…] the demand to use real world applications, […] has not yet been realised widely in German school life."

In Saxony-Anhalt, one federal state in Germany, there is a new math curriculum for secondary schools, the 7th to 13th forms (13 to 19 year olds). There is no special unit for modelling activities but there are many references to problem solving skills.

More than 60% of about 100 pupils (14 to 16 year olds from a grammar school in Magdeburg), when asked to solve word-problems like: *The school's solar panels collect 15kW of energy every hour. How many hours a day must the sun shine in order to collect 180kW?*, "solved" the tasks by manipulating the given numbers without any critical reflection. When questioned afterwards the pupils revealed that they knew about the complexity of the tasks and were also aware of the knowledge that the "simple results" can not be true. In summary, most of the pupils said: "In mathematics I would find these results (because I have to calculate it with the given numbers), but in reality there is a great difference."

However, there is progress; curricula and text books show modifications towards more real world applications and problems. Corresponding learning materials are available and in some federal states of Germany modelling is obligatory in the curricula.

2 Spreadsheets in The Modelling Cycle

In most of the real world problems there is the necessity to use tools like graphic calculators or computers to compute model solutions. Nowadays, pupils have access to modern computer systems, but normally they are not equipped with special modelling software. Thus, the spreadsheet has become a most common and powerful instrument to implement modelling activities especially in lower secondary schools. Spreadsheets are widely available and almost all computers come with a package that includes a spreadsheet program.

The process of modelling is an constant oscillation between various levels of reality. The modelling cycle can be divided in the following phases.

- get to grips with the problem (define the key questions)
- formulate a mathematical model
- generate solutions (from the mathematical model)
- validate the model (and if necessary re-formulate the model until it fits with the real-world-context)

Figure 1 shows the modelling cycle.

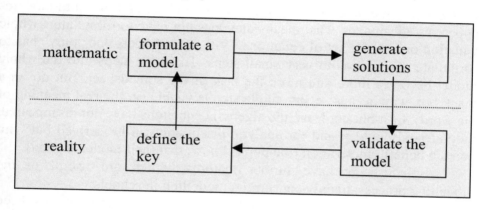

Figure 1: The Modelling Cycle

The process of modelling continues until the validation criteria are accepted by the modeller. In the cycle, one uses different kinds of languages (in "reality" one uses the spoken language and in "mathematics" one uses the mathematical language).

Each of the four parts or steps of the modelling process can be a single component of an task for pupils or can be a topic for discussion in class-rooms as well as the whole process. For example, it is possible to concentrate on the formulation of the model without solving it, or one can give a problem/task and a mathematical solution to the model and the pupils have to find and discuss evaluation criteria.

Pupils in lower secondary schools especially have problems formulating their ideas or assumptions in a mathematical (algebraic) way. It is not easy for them to define variables, formulate equations or inequalities.

The transfer from spoken language into mathematical language (Fig. 1) becomes difficult as well, as it is not easy for the pupils to generate solutions. Often the mathematical model is unsolvable with school skills. (For example problems involving differential equations.) Often difficulties occur and this results in displeasure or failure.

With spreadsheets these effects can be reduced in some fields because it is not immediately necessary to define variables and formulate equations. There is the possibility for a more intuitive or "real-life-use" of mathematical language.

Especially for "modelling beginners" this could be an advantage. Furthermore, the possibility of intuitive use and the splitting into modules appear as advantages. That means one does not have to formulate a whole equation or a whole set of equations. With spreadsheets one "only" has to formulate one's ideas in very small steps. To fit the steps into the whole model becomes more and more the task of the spreadsheet. But not in a black box way because every single step is visible (in the real meaning of the word). Last but not least, the algebraic equations have not disappeared, they are only hidden and for analytic discussion can be derived but with one big benefit: motivation (the pupils know, that their model is good).

Spreadsheets have further features, which are useful in the modelling process, mainly concerning modelling in schools:
- spreadsheets are simple-to-handle tools (without needing advanced computer knowledge)
- spreadsheets allow concentration on the modelling process without dealing with software-skills
- spreadsheets handle large quantities of data
- iterations and recursions are simple to implement
- visualisation of data, relations and functions can be achieved in an easy way
- systematic testing can be carried out(manipulate input-data and observe the resulting output-data)
- spreadsheets provide tools like sum, mean, max, min, etc.
- the table describes the problem, represents the model and often its solution
- spreadsheets are useful in almost every field of real-world problems

In summary of the mentioned features, spreadsheets are an educational tool which are useful to teach efficiently the modelling

process and better still, may be used in "real" real-world problems, not only in schools. Spreadsheets are employable with:
-models which process large quantities of data in an elementary fashion.
-models which are based on iteration and recursion or solved by using systematic testing.
-models which evaluate qualitative and quantitative assumptions.
-models which simulate operations or functional correlations.
-models which deal with visualisations of data or functions.
-models which deal with discrete phenomena.

3 Examples

The following problems and tasks are inspired by the work of Niss et al (1991), Dobner (1997), Houston et al (1997) and Matos et al (2001) and were constructed by the authors to fit in their chosen classroom situation.

All tables were constructed by pupils (14 to 16 year olds from a grammar school) after the teacher introduced the problems and discussed possible ways to solve them. The pupils found their solutions by working together in teams of 2 to 4 persons. Spreadsheets are in common use in this class because of the lessons in computer sciences one year before. Afterwards the solutions are presented by the groups and discussed with the other teams in the classroom.

Example 1

The first example demonstrates the advantage of spreadsheets in the field of model validation. Pupils often do not know whether their model is good or not, especially if there is no precise answer to the given problem. They try to compare it with the other pupils in the class and fail because they only find another solution (mostly different from their own). The well-known Fermi task: *"How many piano tuners exist in Chicago?"* is a simple example of an open problem for the class. In this problem the pupils have to define a criterion to evaluate the model because there is no true answer. (Of course one could try to ask the Trade Corporation and hope they have the true number.) The table shows one way, using a spreadsheet for the calculations, which was created by pupils. At the beginning only the size of the population was given. The table shows the result of 20 minutes working time.

How many piano tuners exist in Chicago?				
			number of pianos	
population	3 000 000			
schools (each 12th is a pupil; 900 pupil each school)	280		280	(1 piano each)
households (4 persons each)	750 000		187 500	(each 4th household)
theatres/ opera houses	50		100	(2 pianos each)
			187 880	total number
piano tuner				
pianos each day	2			
working days a year	300			
frequency per piano (once in ... years)	8			
piano tuners in Chicago	39,1			

Table 1 (translation of a pupil's work on the Fermi task)

The interesting point is the answer of the question: Could this be true? Or better: Could this be probably a good estimated value? Or more mathematical: Is this a good model? Pupils have big problems with the evaluation of their problem solution because they are used to get a 'right' or 'false' answer. They are trained to calculate the right solution, not to get a good solution or a model that fits the real world context.

Spreadsheets provide a possibility to validate the model by parameter variation. Of course in the example mentioned the calculation itself is not so interesting but the assumptions about the input-data. The pupils can check their assumptions by changing (for example doubling) the number of schools or theatres and finding that the number of piano tuners doesn't change. On the other hand, if the pupils divide the number of working days by two the number of piano tuners will double. It shows that the formulation of the given numbers and the model is probably good. And it shows the potentials of the spreadsheet, too. Pupils would lose interest if they had to work out the numbers with paper and pencil (even if they use a calculator). They cannot fail and do miscalculations. They can play with the numbers and verify their ideas very quickly. It is mathematical experimentation and working with a model on a simple level.

Example 2

The second example demonstrates the use of a spreadsheet in a real world problem. It is the so-called "gasoline-problem". In Germany nearly no week passes by without some public discussion about the gasoline price (and its increase). In local newspapers and car magazines there are published tables with price comparisons and many people try to tank up at a low price even if they don't use the nearest gas station but the gas station with the lowest price.

Under these circumstances there occurs the problem: *Is it worth tanking up at another gas station than the nearest?* This open problem could by modelled under different viewpoints. It depends on the understanding of "worth". At a first stage almost all people consider the price as a factor. At a second stage one could think of factors like time or environment and so on. The problem has to be formulated more precisely. How far is it worth going to tank up? Or: In which connection do the factors' price difference, distance between the nearest gas station and the gas station with the lowest price and tank capacity and other factors like time, environment and so on interact? Normally one would now define variables and build up an equation and try to answer these questions. If you use a spreadsheet one could formulate the model on a more intuitive way and be able to look at the problem from different sides. The given numbers are the prices, distances and the gas consumption.

Gas-Problem						
price at station A	2,04	DM		total costs A	81,6	DM
price at station B	2,00	DM		travelling expenses	1,40	DM
gas consumption	0,07	litre/km		gas expenses B	80	DM
tank capacity	40	litre				
distance A to B	5	km		total costs B	81,40	DM

Table 2

Table 2 shows a simple model of the problem only considering the price, distance and tank capacity. But even at this level the pupils can check their assumptions, play with the numbers and answer questions like: How far is it worth travelling at a given price difference? How much do I have to tank? Which influence does the gas consumption have? Comparing this table with the 'equivalent' inequality

$$p_B \cdot c + 2 \cdot d \cdot g \cdot p_B < p_A \cdot c$$

($p_{A/B}$ ... price; c ... tank capacity; d ... distance; g ... gas consumption) the table is much more "real". Of course it is possible to derive the same results by manipulating the inequality, but the table is much more vivid (especially for lower secondary school pupils). After looking at the model at this first level it is possible to re-formulate the model under consideration of the actual position between A and B, remaining gas and on another level time costs and so on. This problem is didactical interesting because it is an example for iterative modelling (every model level fits the reality better).

Example 3

The last example deals with an interesting phenomenon in the bathroom and shows the potential of spreadsheets to handle large quantities of data and models which are based on iteration. The phenomenon is: *If you open the tap the wash-basin fills up to a certain depth (height) and doesn't increase if the outlet is open.* If the pupils consider this problem it is not as easy as it appears at the first moment, because it is obvious that the water stops if there is an equivalent inflow and outflow. But the outflow velocity is not constant and grows non-linearly with the water depth. There are questions about the relations between radius of the outlet and water depth or relations between inflow and water depth.

In the lower secondary school under consideration, there is no possibility to build up a model with differential equations or include effects like swirls and so on. Table 3 shows a model which was formulated by pupils after analysing the problem verbally, especially talking about the dependence of velocity ($v = \sqrt{2 \cdot g \cdot h}$; g ... gravitational pull of the earth; h ...height) and the outflow volume in time ($V_O = \pi \cdot r^2 \cdot v \cdot \Delta t$; r ... radius of the outlet) and the assumption that the basin is a cuboid. The assumptions concerning inflow and base area are made by the whole class before beginning the modelling process.

wash-basin problem					
inflow	10	litre / min	166,7	cm^3 / s	
base area	900	cm^3			
outlet radius	0,7	cm			
time interval	10	s			
time / s	V_{Inflow} / cm^3	$V_{Outflow}$ / cm^3	total volume / cm^3	depth / cm	velocity/cms^{-1}
0,0	0,0	0,0	0,0	0,0	0,0
10,0	1666,7	0,0	1666,7	1,9	60,3
20,0	1666,7	927,9	2405,4	2,7	72,4
30,0	1666,7	1114,7	2957,4	3,3	80,3
⋮	⋮	⋮	⋮	⋮	⋮
620,0	1666,7	1666,6	5377,0	6,0	108,3
630,0	1666,7	1666,7	5377,0	6,0	108,3

Table 3

With this spreadsheet model now the pupils can again play with the numbers and assumptions without any need to do calculations by hand.

4 Conclusion

Spreadsheets provide a powerful, multipurpose tool to teach mathematical modelling. They are useful in almost all kinds of real world problems especially in connection with models, which deal with iterations, recursions, visualisation, simulation and mathematical experimentation. Pupils are almost free from technical problems (model building or model solving) and able to build a model in a very intuitive way.

The modelling process (cycle) is one didactical way of putting more emphasis on real world problems in schools even at lower levels.

If we want our pupils to develop the ability and skill to solve complex real-world problems at the end of their secondary school life, then we have to implement modelling activities in our classroom as early as possible. We have to start at primary school level with uncomplicated but exemplary problems and increase the complexity in the following years, in lower and upper secondary schools, until final examinations. This will achieve one of the most important goals of mathematical education (besides the acquisition of elementary techniques), namely the ability to handle recurring problems from different parts of life with mathematical methods, and to use mathematics to understand real-world-problems and real-world phenomena better and to solve them in an (intelligent) mathematical way.

References

Dobner H-J (1997), 'Mathematisches Modellieren im Schulunterricht', In: Math. Schule.-Päd. Zeitschriftenverlag, Berlin.

Henning H, Keune M (1999) 'Modellbildung beim Aufgabenlösen im Mathematikunterricht' In: PM Praxis der Mathematik, Köln: Aulis Verlag Deubner and Co.

Henning H and Keune M (2001) 'Diskrete Modellbildung und Tabellenkalkulation' In: Der Mathematikunterricht MU 3-2001

Houston SK, Blum W, Huntley ID and Neill NT (1997) 'Teaching and Learning Mathematical Modelling', Chichester: Albion Publishing Ltd (now Horwood Publishing Ltd.).

Matos JF, Blum W, Houston SK, Carreira SP (2001) 'Modelling and Mathematics Education', Chichester: Horwood Publishing Ltd.

Niss M, Blum W and Huntley ID (1991) 'Teaching of Mathematical Modelling and Applications', Chichester: Ellis Horwood.

10

Technology Enriched Classrooms: Some Implications for Teaching Applications and Modelling

Peter Galbraith
University Of Queensland, Australia
p.galbraith@mailbox.uq.edu.au

Merrilyn Goos
University Of Queensland, Australia
m.goos@mailbox.uq.edu.au

Peter Renshaw
University Of Queensland, Australia
p.renshaw@mailbox.uq.edu.au

Vince Geiger
Hillbrook Anglican School and University Of Queensland, Australia
Vincent@gil.com.au

Abstract

We report on a research study that investigated interactions between students, teachers, and technologies in senior secondary mathematics classrooms. Based on data from classroom observation and lesson videotapes, student interviews and questionnaires, we propose four roles for technology in relation to such interactions: master, servant, partner, and extension of self. We illustrate the roles of technology by presenting episodes from lessons that were focused around application tasks. Our analysis shows how different technologies as utilised by individual students, mediated the students' social interactions, and were integrated into the production of mathematical arguments. Implications for the approach to applications and modelling activities are discussed.

1. Background

Much continues to be written about the effectiveness of technology in mathematics learning, and many research studies have set out to examine the effects of technology usage on students' mathematical achievements and attitudes, and their understanding of mathematical concepts (e.g. Adams, 1997). However, the quasi-experimental design of many of these studies is based on the assumption that the same instructional objectives and methods are valid for both pen and paper and technology enhanced tasks. Less is known about how the availability of technology, especially graphical calculators and their peripheral devices, has affected teaching approaches (Penglase and Arnold, 1996).

At the program level one of the most substantial initiatives involving technology use is described by Olsen (1999): She describes how politicians visiting Virginia Tech's Mathematics Emporium, a 58 000 square foot (1.5-acre) computer classroom:

> see a model of institutional productivity; a vision of the future in which machines handle many kinds of undergraduate teaching duties-and universities pay fewer professors to lecture…On weekdays from 9 am to midnight dozens of graduate-student and undergraduate helpers can be observed strolling along the hexagonal pods on which the emporium's computers sit. The helpers are trying to spot the students who are stuck on math problems and need help.

At the individual student level Templer et. al. (1998) challenge such factory production models by pointing to concerns identified by students themselves:

> Having mastered the rudiments, the majority of students "began to hurtle through the work, hell bent on finishing everything in the shortest possible time." The following comment (or a close relative), was noted as occurring frequently "I just don't understand what I'm learning here. I mean all I have to do is ask the machine to solve the problem and it's done. What have I learned?"

Yet another perspective focuses on the learning culture, and on the nature of interactions between the human participants with each other, and with the 'tools" available to assist learning. From this theoretical perspective, that draws on socio-cultural theories of learning (Vygotsky 1978; Wertsch and Rupert, 1993), mathematics teaching and learning are viewed as social and communicative activities that require the formation of a classroom community of learners, where the epistemological values and communicative conventions of the wider mathematical community are progressively appropriated and enacted (Goos, Galbraith and Renshaw, 1999). This approach is predicated on three basic assumptions of socio-cultural theory: (1) Human action is mediated by cultural tools, and is fundamentally transformed in the process. (2) The tools include technical and physical artefacts (including technology) but also concepts, reasoning, symbol systems, modes of argumentation and representation. (3) Learning is achieved by appropriating and effectively using cultural tools currently recognised and validated by the relevant community of practice. Rather than relying on the teacher as an unquestioned authority, students are expected to propose and defend mathematical ideas and conjectures, and to respond thoughtfully to the mathematical arguments of their peers.

When we consider applications and modelling as a case in point, we note first that these pursuits (A and M) reside within the totality of mathematics, and hence we need to keep this wider perspective in mind. Figure 1 provides a simple representation of this relationship. Within the oval representing the totality of mathematical learning contexts, mathematical modelling is displayed as a component of interest, using an abbreviated version of the familiar modelling chart. Those elements that are underlined collectively define that 'subset' of modelling commonly referred to as 'applications'. The black arrows indicate that technology may be involved both in modelling and application work, and in other forms of mathematical activity involving a variety of processes and routines. Furthermore when working with models and applications such processes and routines will frequently be imported as tools to enhance solution processes. The double-heads on the arrows indicate, reciprocally, that mathematical insights gained during modelling activity will be integrated into the store of knowledge for potential use in other mathematical pursuits; and that just as technology may impact on mathematical processing within and outside modelling activity, so learning achieved can provide additional insights to influence how technology is used.

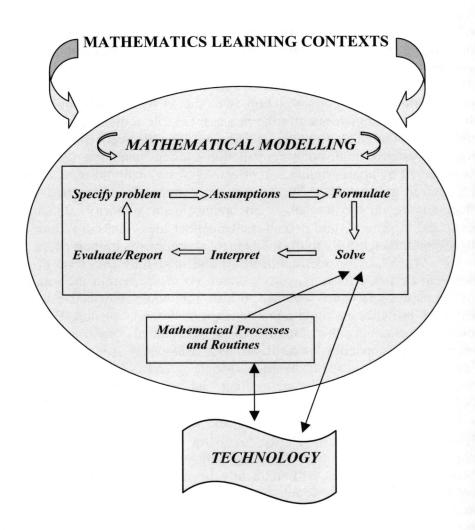

Figure 1. Some technological and mathematical interrelationships

2. The research study

The aim of the study was to investigate ways in which technology enters the mathematical practices in a range of secondary school classrooms. The study was conducted in 5 secondary mathematics classrooms over 3 years, with students 16-17 years old studying under syllabuses that encouraged the use of technology. The teaching content included mathematical applications as well as the wide variety of topics associated with a university preparatory course. Data collection methods included weekly lesson observations and videotaping, interviews with teachers and students, and questionnaires. Since the aim of the study was to investigate students' and teachers' use of technology in specific classroom environments, we employed research methods that drew on ethnographic and case study techniques such as participant observation, interviews, survey instruments, and collection of video and audio taped records. At least one lesson every week was videotaped and observed for each participating classroom, and selected segments of the tapes were transcribed for later analysis. More frequent classroom visits were scheduled if the teacher planned a technology intensive approach to the topic. For example, every lesson was observed and videotaped in a two-week unit of work that introduced some of the mathematics of chaos theory, because in every lesson students were using spreadsheets to investigate numeric iteration processes in a variety of contexts (compound interest, population growth, radioactive decay, approximate methods for solving equations).

This paper draws from data gathered in one of these classrooms, in which the use of technology was integral to all teaching situations, including those involving mathematical applications. Hence in drawing implications for the teaching of modelling and applications, we learn also from the contribution of technology in other learning settings.

3. Pedagogical assumptions

Consistent with our sociocultural perspective, we regard technology as one of several types of cultural tools – sign systems or material artefacts – that not only amplify, but also re-organise, cognitive processes through their integration into the social and discursive practices of a knowledge community (Resnick, Pontecorvo and Säljö, 1997). The amplification effect may be observed when technology simply supplements the range of tools already available in the mathematics classroom, for example, by speeding tedious calculations or verifying results obtained by hand. By contrast, cognitive re-organisation occurs when learners' interaction with technology as a new semiotic system qualitatively transforms their thinking; for example, use of spreadsheets and graphing software can alter the traditional privileging of algebraic over graphical or numerical reasoning. Accordingly, learning becomes a process of appropriating cultural tools that transform the relationships of individuals to tasks as well as to other members of their community.

In reviewing studies that have investigated how and why students use technology we note that alternative emphases are represented. Amongst these, Doerr and Zangor (2000) in an observational case study of two pre-calculus classrooms identified five modes of graphical calculator use from a *mathematical standpoint*: computational tool, transformational tool, data collection and analysis tool, visualising tool, and checking tool. Taking a somewhat different approach, Guin and Trouche (1999) categorised their observations of students using graphic and symbolic calculators into *profiles of behaviour*, in order to understand how students transformed the material tool into an instrument of mathematical thought that re-organised their activity. The nature of this transformation varied according to whether the student displayed a random, mechanical, rational, resourceful, or theoretical behaviour profile in terms of their ability to interpret and co-ordinate calculator results. Our own conceptualisation of technology usage in mathematics classrooms differs from analytical frameworks developed in previous research in that it encompasses interactions between teachers and students, amongst students themselves and between people and technology, in order to investigate how different participation patterns offered opportunities for students to engage constructively and critically with mathematical ideas. That is, while technology may be regarded as a mathematical tool (*amplifies capacity*) or as a transforming tool (*reorganises thinking*), it may also be regarded as a cultural tool (*changes relationships between people, and between people and tasks*). In this paper we have particular interest in the latter.

Our analysis of technology focused classroom interactions is framed by four metaphors we have developed to describe the varying degrees of sophistication with which students work with technology technology as "master", "servant", "partner", and "extension of self". These categories were first tentatively identified while classifying observations made during the first year of the study. Progressive refinement of the categories was guided by further classroom observation and viewing of videotapes by the research team, enabling examples to be documented and issues clarified. We now offer a brief summary of the metaphors.

Technology as Master. The student is subservient to the technology-a relationship induced by technological or mathematical dependence. If the complexity of usage is high, student activity will be confined to those limited operations over which they have competence. If mathematical understanding is absent, the student is reduced to blind consumption of whatever output is generated, irrespective of its accuracy or worth.

Technology as Servant. Here technology is used as a reliable timesaving replacement for mental, or pen and paper computations. The tasks of the mathematics classroom remain essentially the same—but now they are facilitated by a fast mechanical aid. The user 'instructs' the technology as an obedient but 'dumb' assistant in which s/he has confidence.

Technology as Partner. Here rapport has developed between the user and the technology, which is used creatively to increase the power that students have over their learning. Students often appear to interact directly with the technology (e.g. graphical calculator), treating it almost as a human partner that responds to their commands – for example, with error messages that demand investigation. The calculator acts as a surrogate partner as students verbalise their thinking in the process of locating and correcting such errors. Calculator or computer output also provides a stimulus for peer discussion as students cluster together to compare their screens, often holding up graphical calculators side by side or passing them back and forth to neighbours to emphasise a point or compare their working

Technology as an Extension of Self. The highest level of functioning, where users incorporate technological expertise as an integral part of their mathematical repertoire. The partnership between student and technology merges to a single identity, so that rather than existing as a third party technology is used to support mathematical argumentation as naturally as intellectual resources. Students working together may initiate and incorporate a variety of technological resources in the pursuit of the solution to a mathematical problem.

4. Sample lesson episodes

We now proceed to illustrate examples of these modes of functioning using excerpts from a series of classroom episodes (see Appendix for problem contexts).

4.1 Biscuit Extract (technology as servant)

In this first example the teacher used an application context to introduce aspects of matrix operations, which then became part of the students' general mathematical repertoire.

Paul: So then, in this one, you have the frequency here, of packets produced. You know, the different types of packets. **Okay, going back to what I was doing before. Question 4, I did matrices (1) times matrices (2) equal . . .**

Teacher: Do it, then! **(Paul works it out on his TI-92)**

> **Paul:** Equals that, which costs…which,…These are different types of packs here, and up here, is cost in cents. **And I just converted that manually and made the matrices in dollars. Is there any way of doing that? - tricky, you know!** Anyway, I just did that by hand and then created a new matrix called SS5.

In this extract Paul is describing his approach to a part of the problem that has been set up by the teacher to introduce the 'new' concept of *scalar multiplication,* which appears in the guise of converting between dollars and cents. Paul rehearses matrix multiplication using the calculator authoritatively in 'servant' mode, but is only able to incorporate the required transformation by hand. So he then combines 'hand' arithmetic with 'automatic' calculator processing to create a new matrix, noting an unresolved challenge with respect to the calculator. "Is there any way of doing that? - tricky, you know!"

But his inference is that the calculator *ought* to be able to do it, and this is subsequently verified.

4.2 Leontif Extract (technology as partner)

In this second extract a group of students are working together to solve a problem involving an application of newly presented Leontif theory. Although routine for the experienced, it was a 'first' application for the students, and involved identifying and carrying out matrix transformations, as well as translation and interpretation of mathematics in context.

> **Helen:** (to Nerida): What did you get for your inverse?
>
> **Nerida:** (dejected) It tells me there's a dimension error, and I don't know why.
>
> **Edward:** Did you get that? **(passing his calculator to Nerida so she can look at his working)**
>
> **Helen:** **(also showing her calculator working to Nerida)** It should be that.
>
> **Nerida:** **(comparing her working with the other two screens)** That's what I had!
>
> **Helen:** So then you ...
>
> **Nerida:** **(puzzled, comparing screens with Edward)** Is that what you have? It's exactly the same as mine!
>
> **Helen:** Yeah, and you times that by 200, 200, 200 *down* (referring to demand matrix)
>
> **Nerida:** (sudden insight) Oh hang on ... That should be *four* ... Oh God!
>
> **Helen:** What did you do?
>
> **Nerida:** I didn't do four 200s!
>
> **Helen:** Oh you big dork! (Nerida and Helen laugh) You've only got three 200s! (referring to number of entries in demand matrix – there should be four rows, not three)
>
> **Nerida:** **(chastened) God I'm a moron! (talking to her calculator as she presses buttons)** Second, quit. Now.. (re-does the calculation). "Asks Helen"- Did you get that?

Helen:	(inspects Nerida's screen) Yeah! (Nerida jumps up from her seat in delight)
Edward:	(to Nerida) Look at mine.
Nerida:	(goes over to Edward) Did you get this? (Edward holds both calculators up side by side, compares his screen with Nerida's. Nerida pulls his hair when he deliberately hesitates in replying.)
Edward:	(with cheeky grin) Yes!
Nerida:	Thank you!

In this extract there are three human collaborators, and three calculators that are integrated as key players in resolving the identified problem. Dialogue moves easily between mathematics and technology. The bold dialogue highlights the team or partnership approach in which human and calculator contributions are seamlessly merged.

4.3 Chaos Extract (technology as extension of self)
The activity within this task extended over three class periods. Discussion in this case involves selections from the whole, and is not appropriately represented by a single piece of dialogue. The teacher (fourth author of this paper) had chosen to introduce students to iteration as one of the central ideas of chaos theory. This topic was presented as a teacher-prepared booklet containing a series of examples and tasks for students to work through at their own pace as unrehearsed challenges.

One such task involved using iterative methods to find approximate roots of equations such as $x^3 - 8x - 8 = 0$. The equation may be expressed in the form $x = F(x)$, and a first approximation to the solution is obtained by estimating the point of intersection of the curves $y = x$ and $y = F(x)$.

Using a spreadsheet the arrangement $x = \dfrac{x^3}{8} - 1$ **yields only one of the three roots**

(–1.236). (To find the other roots of this cubic equation (–2 and 3.236), it is necessary to investigate other rearrangements and a range of initial values.)
One student group (Hayley, Nerida, Sally, David) used graph-plotting software to draw graphs for the LHS and RHS of the above-three intersection points are clearly visible. They zoom in on only one intersection point to find the *x*-coordinate, and obtain an approximate value of 3.24, which they accept as "the" solution – there is no attempt to explore other intersections. They then move on to the next problem although Nerida remains puzzled as to why only one solution was obtained "Why did the solver only pick up one?"
At the start of the next lesson the students return to the cubic problem, this time using a spreadsheet.

They enter a formula equivalent to their original rearrangement of the equation ($x = \dfrac{x^3}{8} - 1$) and "fill down" the columns until the values converge. However, their answer, -1.23 does not match the graphical result obtained earlier:

Sally: But we got 3.24!

Hayley reminds the group that there were three intersection points visible on the graph, and suggests they might find the other two solutions if they continue scrolling down their spreadsheet. When this is not successful they call the teacher over and request clarification as to how the spreadsheet works. He re-focuses the group on the important elements of the task, and issues a challenge:

Teacher: Is it possible to use the spreadsheet to get all three solutions?

The students find that trying different initial values makes no difference to their position: the spreadsheet values either converge on -1.23 or become increasingly large. David reproduces the graph previously plotted on the computer with the aid of the TI-83, thus enabling the graph and spreadsheet to be viewed simultaneously, saying "it's quicker than multi-tasking". Nerida concurs, arguing that "Otherwise we'd have to swap around (i.e. between spreadsheet and graphing program) using the computer and it takes ages".

They continue trying different initial values, to no avail, then, following a suggestion from the teacher disperse to consult with other groups, and discover two other ways of rearranging the equation: $x = \sqrt[3]{8x + 8}$ **and** $x = \dfrac{8x + 8}{x^2}$. These give the "missing" spreadsheet solutions of 3.24 and -2 respectively.

On re-convening as a group, the students pieced together the information they obtained, set up the relevant spreadsheets, compared with the graphical representation, and confirmed they had found all three solutions. This resulted in some excitement as no other group has managed to do this. The approach by these students differs in kind from that illustrated in the Leontif example. There the students found ways to explore and test results by collaborative use of a particular medium - a graphical calculator. Here the students exercise judgement in the choice of medium, having regard to the properties of each and their suitability, either alone or in combination, to address the task at hand. Operating in this mode is not meant to imply that success is always assured. It means that students move within and between technologies (here spreadsheet, graphical calculator, and graphing software program), with the same confidence and insight that good problem solvers display as they access and combine a variety of intellectual skills in pursuit of solutions to problems. Ideally, the former complements and extends the latter.

5. Implications for learning and teaching

The classroom episodes presented above illustrate how students' task performance was shaped by the tools available to them (graphing software, spreadsheet, graphical calculator), by the sociocultural context of the classroom, and by their approaches to using the technology. Integral to the outcomes were the teacher's actions in orchestrating students' interaction with the task, the technology, and their peers. The relationship between technology usage and teaching/learning environments is not one of simple cause and effect. The four metaphors of "master", "servant", "partner", and "extension of self" are intended to capture some of the different ways in which technology enters into the mathematical practices of secondary school classrooms. Included below are sample responses obtained when students were invited to reflect on their own approach and skills with technology.

Master (M): because I often don't understand how to use every specific function of the technology, thereby limiting my use of such technology. I often don't know if I've used it correctly and as a consequence I can't be sure if my answer is correct or not.
I think I'm between *master* and *servant*. I tell the calculator what to do sometimes but only stick to what I know usually. I don't know exactly what it allows me to do, and if I haven't been taught, I won't look for it.
Servant (S): because I do not have enough knowledge of technology to be able to investigate new concepts. However I do regularly use it for *familiar* tasks purely as a time saver and to verify and check my answers.
Partner (P): Because my calculator has become my best friend. His name is Wilbur. Me and Wilbur go on fantastical adventures together through Maths land. I don't know what I'd do without him. I love you Wilbur.
Extension of Self (ES): Because my calculator is practically a part of myself. It's like my 3rd brain. I use it whenever it can help me do anything faster.

We have also observed that students do not necessarily remain attached to a single mode of working with technology-for example sometimes 'servant mode' suffices in routine situations even though students are capable of more sophisticated use.

6. Conclusion

Students' task performance was shaped on the one hand by *tools* (graphing software, spreadsheet, graphical calculator); and on the other by the sociocultural context of the classroom (teacher beliefs and actions in selecting and orchestrating interaction with tasks, technology, peers, and forms of mathematical argumentation and problem solving). Technology, as a cultural tool, re-organises social interactions, changes the way that knowledge is produced, shared and tested, and is integral to the mathematical practice of specific learning environments. Mathematical applications may serve more than one purpose. Of course a fundamental goal remains the helping of students to apply their mathematics to realistic situations. However as in the project described here, applications may also be introduced to provide contexts for the introduction and consolidation of new concepts and skills, which have relevance beyond the specifics of applications and modelling (Figure 1). While this has always been possible in theory, powerful technology now makes this a more practical option as a teaching approach. The different levels of sophistication with which students operate with technology cast a questioning light on assumptions made about the impact of technology on the teaching of modelling and applications. Because the nature of student activity, both individually and collaboratively, can be profoundly changed by the availability of technological devices, it is no longer adequate to consider implications simply in terms of the more powerful mathematical computations that can be achieved. That is, much more is implied than simply being able to set problems requiring more technical mathematical solutions. Such a restricted view would amount to treating 'technology as servant' as the defining metaphor, and we have seen that technology use, and the creative potential of students extends well beyond this routine though important level. In particular implications for collaborative group activity in modelling and applications work are profound.

References

Adams T (1997) 'Addressing students' difficulties with the concept of function: Applying graphing calculators and a model of conceptual change' *Focus on Learning Problems in Mathematics* 19(2), 43-57.

Doerr H M and Zangor R (2000) 'Creating meaning for and with the graphing calculator' *Educational Studies in Mathematics* 41, 143-163.

Forman E A (1996) 'Learning mathematics as participation in classroom practice: Implications of sociocultural theory for educational reform' in Steffe L et al (Eds) *Theories of mathematical learning,* Mahwah NJ: Erlbaum, 115-130.

Goos M, Galbraith P and Renshaw P (1999) 'Establishing a community of practice in a secondary mathematics classroom' in Burton L (Ed) *Learning mathematics: From hierarchies to networks,* London: Falmer Press, 36-61.

Guin D and Trouche L (1999) 'The complex process of converting tools into mathematical instruments: The case of calculators' *International Journal of Computers in Mathematical Learning* 3, 195-227.

Olsen F (1999) 'The promise and problems of a new way of teaching math' *The Chronicle of Higher Education* 46 (7), 31-35.

Resnick L, Pontecorvo C and Säljö R.(1997) 'Discourse, tools, and reasoning' in Resnick L et al (Eds) *Discourse, tools, and reasoning: Essays on situated cognition,* Berlin: Springer-Verlag, 1-20.

Templer R., Klug D and Gould I (1998) 'Mathematics Laboratories for Science Undergraduates' in Hoyles C et al (Eds) *Rethinking the Mathematics Curriculum,* London: Falmer Press, 140-154.

Penglase M and Arnold S (1996) 'The graphics calculator in mathematics education: A critical review of recent research' *Mathematics Education Research Journal* 8, 58-90.

Vygotsky L S (1978) 'Mind in Society', Cambridge MA: Harvard University Press.

Wertsch J V and Rupert L J (1993) 'The authority of cultural tools in a sociocultural approach to mediated agency' *Cognition and Instruction* 11, 227-239.

APPENDIX: SAMPLE APPLICATIONS

1. Biscuits. In planning for Christmas the biscuit factory decided to release three different packets.

Holly Dream (HD): contents 8 chocolate, 12 thin sweet, 6 shortbread, and 4 rich cream.
Sleigh Ringers (SR): contents 12 chocolate, 15 shortbread, and 5 rich cream.
Reindeer Extras (RE): contents 14 chocolate, 16 thin sweet, 12 shortbread, and 8 rich cream.
Manufacturing costs (cents): chocolate 8, thin sweet 4, shortbread 7, rich cream 10.
Packets will be sold to shops for $2.10, $2.95, and $3.70 respectively. Recommended retail prices will be $2.49, $3.49, and $4.49 respectively. Numbers of packets to be manufactured (millions) will be: HD (2), SR (1.5), RE (1).

In your workbook answer these questions:

(1)   Show the contents of the three proposed packets in a matrix with 3 rows and 4 columns.
(2)   Show the manufacturing costs per biscuit in a column matrix.
(3)   Show the numbers of each packet to be manufactured in a row matrix.
(4)   Use the matrices in (1), (2), and (3) to find the cost to the manufacturer.
(5)   Show the prices at which each packet will be sold to shops in a column matrix.
(6)   Use your matrices in (3) and (5) to find the total amount the factory charges the shops.
(7)   Calculate the total amount which consumers will pay for the biscuits.

2. Leontif: An economy with the four sectors manufacturing, petroleum, transportation, and hydroelectric power has the following technology matrix:

$$T = \begin{pmatrix} 0.15 & 0.18 & 0.3 & 0.1 \\ 0.22 & 0.12 & 0.37 & 0 \\ 0.09 & 0.3 & 0.11 & 0 \\ 0.27 & 0.05 & 0.07 & 0.1 \end{pmatrix}$$

Find the production matrix if all the entries in the demand matrix are 200.

3. Chaos Unit content: Here students investigate introductory concepts in the recently developed mathematical fields of Chaos Theory and Fractal Geometry. Subject matter includes both skill development and applications such as: basic notions of recursive functions and iterative processes; exponential growth and decay, the growth equation, stability of systems modelled by differential equations, strange attractors.

Learning Experiences include; use of computer spreadsheeting software to investigate iterative processes in a variety of contexts, e.g. compound interest; uninhibited growth of animal populations; investigating a formula as a model for population growth and decay; investigating the values of 'a' which allow 2 and 3 cycles in the Verhulst equation; using computer software to develop fractal images.

11

Choosing and Using Technology for Secondary Mathematical Modelling Tasks – Choosing the Right Peg for the Right Hole

Vince Geiger
Hillbrook Anglican School
vincent@gil.com.au

Peter Galbraith
University of Queensland
p.galbraith@mailbox.uq.edu.au

Peter Renshaw
University of Queensland
p.renshaw@mailbox.uq.edu.au

Merrilyn Goos
University of Queensland
m.goos@mailbox.uq.edu.au

Abstract

This paper investigates students' preferences in relation to a variety of technologies in a classroom environment where the use of computer and/or graphing calculator technology is strongly encouraged and the teaching/learning program is rich in applicable mathematics. The choices made by students in relation to the appropriateness of a particular technology for specific mathematical tasks are investigated. Factors such as familiarity with a device, available features and display type were identified as well as an indication that students did not distinguish between applications and other types of mathematical tasks when making choices about the appropriateness of a technology.

1. Introduction

It is likely that by 2003 there will be assumed access to mathematically enabled technologies for both teaching, learning and assessment in seven out of Australia's eight states and territories. At the same time the capacity to make use of mathematics in life-related as well as purely mathematical contexts continues to be an important objective in secondary school mathematics curriculum frameworks and study designs (for example Queensland Board of Senior Secondary School Studies, 1992). While there is now a significant body of knowledge in relation to the use of technology in the teaching, learning and assessment of mathematics in general, there is also a growing literature base that examines the potential of technology to enhance student learning in applied contexts (eg Huntley, 1991, Clements, 1991, Geiger, 1999).

This paper seeks to extend knowledge in this domain through the analysis of interviews that targeted students' preferences when a choice of graphics calculators, with different levels of functionality, are available. A number of classes of preference emerge from the analysis.

2. Literature

While many of the predictions of the impact of learning technologies have focused on enhanced student learning outcomes such as concept development (eg, Vonder Embse, 1992; Jones and Lipson, 1993), and varied approaches to instruction (eg, Dance et al, 1992), others have suggested that the most significant changes will be related to the ways students and teachers will interact in mathematics classrooms (Burrill, 1992; Geiger and Goos, 1996; Galbraith et al, 1999). Overall findings, however, concerning the "value added" to students' learning through the use of technology have been somewhat inconclusive (Kuchler, 1998, Lesmeister, 1996, Maldonado, 1998; Penglase and Arnold, 1996), although many studies have reported that the use of technology has a positive effect on students' understanding of function and graphing concepts, spatial visualisation skills, and connections between mathematical ideas (Portafoglio, 1998). Further, Kutzler (1999) has argued that mathematically enabled technologies have the potential to scaffold student learning in such a way that gaps in prior learning can be managed so that they do not interfere with the acquisition of new mathematical ideas and concepts.

While the use of technology is an integral part of the modelling process for professional mathematicians, it is a relatively new initiative within secondary schools. The developing incorporation of technology into school mathematics programs, however, has provided some with the opportunity to trial the use of various technologies from graphics calculators to computer algebra systems. Brown (1998) provides an illustration of how graphics calculators can be utilised as tools for the construction of mathematical models to fit life-related data, and that these models can be explored through symbolic, graphical and numeric representations. This is complemented by Henn's (1998) observation that computer algebra systems supply similar potential but with the added benefit of providing the extra algebraic dimension to a modelling investigation. It is unlikely, however, that the true potential for the use of technology in the teaching and learning of applications and modelling will be realised until there is a more widespread implementation of technology in mathematics classrooms.

3. Contexts of the Study

This study took place within a teaching/learning situation in which a number of contexts add colour and texture to the events which take place. A description of these contexts follows.

3.1 Curriculum Context

The study took place within a secondary mathematics classroom in the state of Queensland, Australia. Secondary schools in Queensland work within a school based externally moderated assessment system. Although the curriculum framework for each school must be based on state-wide subject syllabuses these syllabuses are only mildly prescriptive and as a result schools enjoy a high level of flexibility in the design and implementation of curriculum and assessment programs.

There are, however, a number of non-negotiable aspects of each syllabus. For mathematics this includes the mandatory teaching, learning and assessing of mathematics within application based contexts as well as purely mathematical contexts. These applications must appear in contexts that are both familiar and non-routine to the student. Thus the assessment of a student's capacity to deal with applications occurs in contexts that are considered well rehearsed and also in situations where students must make use of problem solving capabilities to deal with applications set in novel contexts. Applications tasks must also vary in relation to the complexity of the task. Two problems are presented below as typical examples of applications problems.

The problem above would be considered a relatively straightforward application of mandatory content within the syllabus. The familiarity of the task is, of course, dependent on the prior learning experience of the student.

Example 1

A tennis player has attempted to mix up the direction of her serves but in fact the following pattern has emerged.

- If she has just served to her opponent's forehand side then on 0.85 of all occasions her next serve will be on the backhand side.
- If she has served to her opponent's backhand side then on 0.65 of all occasions her next serve will be on the backhand side.

If this pattern is maintained over a lengthy period of time, such as a season, what proportions of serves will go to the forehand and the backhand during this period?

While the problem below represents a purely mathematical context, the Fibonacci sequence was introduced to students via a discussion of the breeding patterns of rabbits. This problem illustrates how, within this classroom, ideas are introduced through applications of mathematics but are often extended to include purely mathematical perspectives on the same topic.

Example 2
The Fibonacci sequence can be expressed recursively as:
$F_0 = 0, F_1 = 1; Fn = F_{n-1} + F_{n-2}$
The first few terms in this sequence appear below:

n	0	1	2	3	4	5
F_n	0	1	1	2	3	5

Use this as a basis to prove the following relationship.

For $A = \begin{vmatrix} 1 & 1 \\ 1 & 0 \end{vmatrix}$ $A^n = \begin{vmatrix} F_{n+1} & F_n \\ F_n & F_{n-1} \end{vmatrix}$ where F_n is a term in the

Fibonacci sequence.

This is an example of a problem used to assess a student's capacity to link different areas of mathematics in a single application task. This would be considered a problem of relatively high complexity and, provided there has been no direct tuition in relation to the task, unfamiliar territory.

3.2 Classroom Context and Data Source

This paper seeks to describe one aspect of a larger two year study which aims to investigate the teaching/learning practices of students and teachers in senior secondary school mathematics courses in relation to the use of technology. The issues that this paper describes emerged from students' use of technology when attempting to solve problems, directly or indirectly related to applications, within everyday classroom practice. The students were involved in their final two years of study in a course designed for tertiary bound students with aspirations for further specialist studies in mathematics or related disciplines. The 15 students (5 female, 10 male) were typically 16 – 17 years old and by nature of their choice of subjects generally displayed a positive disposition to the study of mathematics.

The data sources for this project consisted of: audio and video records of classroom observations; individual, group and whole class structured and semi-structured interviews; and questionnaires. The data on which this particular paper is based was drawn from three 45 minute lessons spaced over a two week period. These lessons were video taped.

In the first lesson students worked through a combinatorics problem that was unfamiliar in nature. It was noted that students made use of the available technology in a number of different ways and so the project team decided to follow up with a whole class interview using the parts of the video of the previous lesson for stimulated recall. Approximately one week after this session, students undertook a course work focused assessment task. During this assessment task it was again observed, by the classroom teacher, that students were using different calculators and using them in different ways. A second whole class interview was conducted during the next available lesson in relation to the students' use of calculators during assessment.

3.3 Technology Context

There are no restrictions on the form of educational technology that students are allowed to employ in the learning, teaching or assessment of mathematics in the school based system of instruction and assessment within which this study is embedded. Students and teachers are permitted to use whatever technology they can access. This does not mean, however, that there is wide use of technology in mathematics classrooms across the state; rather the degree of use varies from school to school and depends largely on the expertise and interest of the teaching staff within mathematics departments. The decision as to whether or not to use technology is left to the discretion of each school, and usually each subject department within each school.

The particular school mathematics department from which the data for this study are sourced had made the decision some years previously that the learning of mathematics could be greatly enhanced through the use of technology. As a result it had become a matter of policy that technology be integrated into student learning experiences whenever it was appropriate and possible. The type of technology utilised ranged from computers through to graphics calculators. The student group that is the focus of this paper made use of graphics calculators in particular. There were two types of calculator used. These are listed below with a brief outline of the features available on each:

TI-92 Calculators (Texas Instruments)
- Graphical, numeric representations of functions
- Statistical capabilities
- CAS capabilities - including matrix editor and complex number manipulation
- Dynamic geometry module

TI-83 Calculators (Texas Instruments)
- Graphical, numeric representations of functions
- Statistical capabilities

Students in this group also displayed a positive predisposition to the use of technology (Galbraith at el, 2001).

4. Teaching Episodes and Observations

4.1 A Problem in Combinatorics

In general, new ideas in mathematics were introduced to students through a practical context. The fundamental ideas of combinatorics, for example, were introduced to students through discussions on the number of possibilities for the choice of a committee of set size from a larger number of nominations. These discussions, however, eventually lead to a more formal and abstract treatment of the topic. The first in the series of lessons that are the focus of this paper had moved from a discussion about combinatorics in context to attempting to solve the problem below:

$$^{n}C_3 = \frac{5}{18}\,^{n}C_5 \qquad Find \ n$$

It was found that students used both types of calculators available to assist them to solve the problem but also took different approaches to how they used the calculators.

4.2 Student Revelations

The next day students were shown video segments of the whole class working on the combinatorics problem from the previous lesson. The first question they were asked was why they would choose one calculator type over another (TI-83 vs TI-92). One student argued that you should only use technology as powerful as is needed for the task at hand. Other students commented that they preferred to use one calculator over another simply because they used it more often (eg in other courses) and were more familiar with its operation.

Another student commented that the TI-92 calculator offered more direct access to the commands than the TI-83 which was more menu driven. This indicated that the student had developed a level of familiarity and expertise with the device, that his focus was now on working with the mathematical task, and the calculator was a useful tool to be used as efficiently as possible.

Teacher. But Stephen, why would you have picked the 83 rather than the 92

S1. Because I don't think there is anything more the 92 could have done that the 83 couldn't have...

Teacher. So Stephen, the idea, what you are telling me is that unless you really need the facilities of a 92 you'd prefer to use the 83?

S1. Yeah.

Teacher. Is there any reason why...that you prefer to use it?

S1. Well it types faster.

Other students

S2. 'cause we use it (TI-83) more often, it's easier to know where all the functions are

S3 We're used to it.

Teacher. Whoa, whoa...let's slow down.... So you think the nice thing about the 92 ...'cause you (talking to Student D) tend to use it as a first preference, don't you, is that because you know the command you can just type in and do things straight away, whereas on the 83, you have to go through the menus to get the commands?

Student agrees.

Other students indicated that the choice of calculator depended on the task and the particular device's suitability to that task.

G. Well it depends on what kind of work you're doing...with permutations and combinations, I find it easier on this (signalling TI-83) but with complex numbers and things (another student adds matrices) yeah, matrices - it's so much easier on the 92.

Some students were even prepared to use two calculators at once to satisfy a desire to use, in their view, the technology was most suited to a task or sub-task.

S4. Whenever I was doing normal calculations, I'd use the little one because I could look at my matrix and do the calculators at the same time, that sort of thing.

Teacher. You're using both at once because...

S4. By then, you could see the matrix really clearly on the big calculator

Teacher. You use that as just a way of looking at it, but you do all the manipulation on the other one.

S4. Well I like using the big one for complicated stuff like "i" and stuff like that, but the little one is just easier to use when you are doing all that little basic stuff.

Students also commented on differences between the way the calculators displayed certain inputs and outputs. They indicated a preference for a visual display that might be considered closest to how mathematics is written with a pad and pen, or when properly type set. They found this more useful for checking their inputs and looking for mistakes.

Students were also asked to comment on their choice of methods when solving the combinatorics problem that had been the focus of the initial lesson. Students commented they had simply made use of the calculator to implement whatever method of solution had occurred to them at the time. For some this meant a "trial and error" approach, for others, perhaps more familiar with the "equation solver" capacities of the calculator, this meant the use of a numeric approach to solution. The teacher also inquired what an absent student was doing on the video. Students working with her replied as below:

S5. It (the TI-92) gives you what you typed in. It actually writes out the matrices you are multiplying together instead of giving just "a" or"b".

Teacher. Okay, so you can see it....you like seeing the matrix, not just a symbol for it?

S6. Visually this (the TI-92) is a lot better.

Second interview

S7. Um, I usually pick the big one (TI-92) when we've had matrix problems, where we had big numbers...you could see the whole thing.

S8. Because it's easier to see and you can find the mistakes.

Teacher. You know what she (N) was doing E.

S9. She was expanding

Teacher. So she (E) wasn't solving like D, she was expanding using it like an "Algebra Assistant". She was using it as a way of checking her algebra.

S9. She used it to *do* her algebra.

This student appears to have used the calculator to compensate for a lack of capacity to perform algebraic operations. Her understanding was at a level of sophistication where she knew an algebraic approach was appropriate but did not have the capacity or lacked the confidence to follow this up without assistance. The CAS facility of the calculator had provided the scaffolding she needed to continue with her chosen approach. Given her need for this support it seems unlikely she could have completed this problem using this method if she did not have the "assistance" offered by the calculator available.

4.3 The Role of Technology in the Process of Modelling

An interesting observation was an absence of comment about the use of technology in relation to application tasks. While it was to be expected that students confined their first comments to matters of a purely mathematical or technical nature because they were dealing with a combinatorics problem that was without any life related context, it is a little surprising that no comments were recorded that related to the use of technology in applied tasks. Students had worked with mathematical applications within each topic area of the course. As commented earlier topics were generally introduced through a life related mathematical problem. Further life related tasks were integrated into coursework. Students had also participated, very recently, in an assessment program that by specification maintained a balance between life related and purely mathematical questions and tasks. Yet no student commented on anything that related application of mathematics to technology.

Explaining the absence of an expected observation is of course always a matter of conjecture but perhaps the reason that students failed to relate technology to the modelling process is that it is at the point of abstraction in this process that they see technology being applied. This point of interaction is represented in Figure 1 and is consistent with the generally accepted process of mathematical modelling. The absence of comment by students in relation to technology and the processes of modelling and recontextualisation may indicate they believe technology is a tool to be used to interact with mathematical ideas only after a mathematical model is developed rather than as a tool for exploration and development of a model or its validation as a reliable representation of a life related situation.

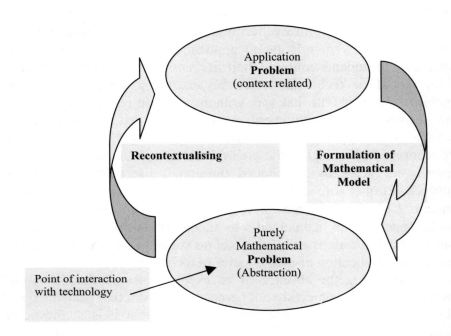

Figure 1. Technological and modelling interaction

5. Discussion

A number of observations are emergent from these student comments. Familiarity with a technology and confidence in using technology appears to be a significant determinant in a student's choice of technology - in this case a choice between two different graphics calculators. It should be noted, however, that this is not the only influence.

It was clear that students had the capacity to make a decision on the choice of technology based on a device's suitability to a task. So, for example, a student might decide to use a TI-83 for arithmetic work but change to a TI-92 in order to work with matrices.

Students also reflected on the differences between visual displays when making a choice between calculators. A number of students commented on the importance of seeing the whole of their input or expected outputs while working with a calculator. They felt less comfortable in working with a problem or detecting errors when some part of the input or output was off screen because of the length of the input or output. Further, students also expressed a preference for displays that represented mathematics in a similar fashion to the way they wrote it with a pen on paper, or how it appeared in a book. Clearly they found unfamiliar styles of symbolic representation at best distracting and at worst less intelligible. Thus the "visual" aspect of students' interaction with mathematics and technology should not be underestimated.

It was observed that at least one student made use of the CAS capabilities of one the calculators to support her use of an algebraic approach to solving the combinatorics problem. This illustrates the potential that exists for CAS technology to provide the scaffolding needed by students who are less proficient with algebra to access other aspects of mathematics. Further work is needed in this area to determine the potential support this technology offers students, who are less proficient in terms of traditional algebraic capabilities, to find success in applications and modelling courses.

Finally, it was noted that students did not distinguish between applications or purely mathematical contexts when discussing use and preferences in relation to technology - all discussions were confined to mathematical ideas at the level of the abstract. While it is a matter of conjecture that students perceive the use of technology, in the process of mathematical modelling, to be only appropriate during the abstraction phase, this observation serves as a warning about assumptions that might be made about how students will use technology in solving application problems. It would seem that students do not readily recognise the potential offered by technology to support them through other phases of the modelling process without direct attention to this issue. The implication, for teaching and learning, is that students may need directed support to realise ways in which technology can be used to assist in the construction, testing and validation of models. This support would necessarily include the provision of further tasks which provided opportunity for students to engage in the process elements of mathematical modelling - exploration, model building and validation. Such tasks would need to be used both as the means for a teacher to model these processes, and also as catalysts for students' engagement in the modelling process.

6. Conclusion

Students indicated preferences for the use of different facilities available on particular technologies. These preferences can be task related but are also dependent on their degree of experience with a particular technology. Such predispositions potentially influence their approach to learning within technologically rich environments and so may affect their capacity to succeed in classroom cultures where the use of technology is an expected norm. This means, for example, that assessing the performance of applications and modelling skills might be confounded when technology use is essential to the solution process. Such concerns may be further exacerbated if a broader range of technologies is available than was the case in this study, particularly if the available technologies provide greater facility and capacity than the specific technologies used by students here.

In this educational setting mathematical applications are introduced throughout the program, and integrated with other content that represents formal structures, and pure mathematical material. It is interesting that student comments regarding their preferences for technology use also do not make distinctions between pure and applied mathematical activity, but focus on the immediate needs of the job in hand. The power of these student preferences in relation to learning styles and approaches are strongly influential, and need further delineation with respect to the teaching and learning of all mathematics, including applications and mathematical modelling.

References

Brown R (1998) 'Mathematical modelling and current events using hand held graphing technology' In Galbraith P et al (Eds), *Mathematical modelling: teaching and assessment in a technology-rich world* Chichester: Horwood, 85 – 94.

Burrill G (1992) 'The graphing calculator: A tool for change' In Fey J and Hirsch C *Calculators in mathematics education* Reston, Virginia; NCTM, 14 –22.

Clements R (1991) 'The integrated microcomputer software environment for the support of mathematical-modelling teaching' In Niss M et al (Eds) *Teaching of mathematical modelling and applications* Chichester: Horwood, 351-362.

Dance R, Nelson J, Jeffers Z and Reinthaler J (1992) 'Using graphing calculators to investigate a graphing calculator model' In Fey J and Hirsch C (Eds) *Calculators in mathematics education* Reston Virginia; NCTM, 120-130.

Galbraith P, Renshaw P, Goos M and Geiger V (1999) 'Technology, mathematics, and people: Interactions in a community of practice' In Truran J and Truran K (Eds.), *Making the difference* (Proceedings of the 22nd Annual Conference of the Mathematics Education Research Group of Australasia) Sydney: MERGA, 223-230.

Galbraith P, Renshaw P, Goos M and Geiger V (2001) 'Integrating technology in mathematics learning: What some students say' In Bobis J et al (Eds.) *Numeracy and beyond* (Proceedings of the 24nd Annual Conference of the Mathematics Education Research Group of Australasia) Sydney: MERGA, 225-232

Geiger V and Goos M (1996) 'Number plugging or problem solving? Using technology to support collaborative learning' In Clarkson P (Ed) *Technology and mathematics education* (Proceedings of the Nineteenth Annual Conference of the Mathematics Education Research Group of Australasia) Melbourne: MERGA, 10-19.

Geiger V (1999) 'Students using of technology in applications - partners, pasters or plaves?' (Paper presented in VC99 - Investigating the investigative; some issues and themes in contemporary open ended approaches to mathematics in schools, the second virtual conference) The Australian Association of Mathematics Teachers Inc. September: AAMT.

Henn H (1998) 'The impact of computer algebra systems on modelling activities' In Galbraith P et al (Eds) *Mathematical modelling: Teaching and assessment in a technology-rich world* Chichester: Horwood, 115-124.

Huntley I (1991) 'Computing mathematics - a new direction for applied mathematics?' In Niss M et al (Eds) *Teaching of mathematical modelling and applications* Chichester: Horwood, 46-62.

Jones P and Lipson K (1993) 'Determining the educational potential of computer based strategies for developing an understanding of sampling distributions' In Atweh B et al (Eds) *Contexts in Mathematics Education* (Proceedings of the sixteenth annual conference of the Mathematics Education Research Group of Australasia) Brisbane: MERGA, 355-360.

Kuchler JM (1998) 'The effectiveness of using computers to teach secondary school (Grades 6-12) mathematics: A meta-analysis' (Unpublished Doctoral dissertation, University of Lowell). Dissertation Abstracts International, 59, 10A, 3764.

Kutzler B (1999) 'The algebraic calculator as a pedagogical tool for teaching mathematics' (Paper presented in VC99 - Investigating the investigative; some issues and themes in contemporary open ended approaches to mathematics in schools, the second virtual conference) The Australian Association of Mathematics Teachers Inc. September: AAMT.

Lesmeister LM (1996) 'The effect of graphing calculators on secondary mathematics achievement' (Unpublished MS thesis, University of Houston). Dissertation Abstracts International, 35, 01, 39.

Maldonado AR (1998) 'Conversations with Hypatia: The use of computers and graphing calculators in the formulation of mathematical arguments in college calculus' (Unpublished Doctoral dissertation, The University of Texas). Dissertation Abstracts International, 59, 06A, 1955.

Penglase M and Arnold S (1996) 'The graphics calculator in mathematics education: A critical review of recent research' *Mathematics Education Research Journal*, 8, 58-90.

Portafoglio A (1998) 'The effects of pair collaboration in community college computer calculus laboratories'. (Unpublished Doctoral dissertation, Columbia University Teachers College). Dissertation Abstracts International, 59, 07A, 2407.

Queensland Board of Senior Secondary School Studies (1992) 'Senior Syllabus in Mathematics C' Brisbane: Board of Senior Secondary School Studies.

Vonder Embse C (1992) 'Concept development and problem solving using graphing calculators in the middle school' In Fey J and Hirsch C *Calculators in mathematics education* Reston, Virginia; NCTM, 65-78.

Section D

Models for Use in Teaching

12

Groups, Symmetry and Symmetry Breaking

Albert Fässler
Hochschule für Technik und Architektur HTA, Biel/Bienne, Switzerland
Albert.Faessler@hta-bi.bfh.ch

Abstract

This paper deals with the teaching of elementary group theory in a way that appeals to students and motivates them to work in an interdisciplinary and application oriented setting. Textbooks often offer purely algebraic and abstract definitions that are difficult to grasp, because the students can neither make sense of them nor make them a part of their lives. Nevertheless, symmetry considerations are a natural way to represent groups. Group theory, on the other hand, is an indispensable tool for studying symmetry. Furthermore, knowing symmetry breaking allows us to explain or make hypotheses about non-linear phenomena in various fields, like biology or astronomy. Symmetry breaking is a modern concept and an attractive field of research in applied mathematics. The fact that it occurs in many everyday situations makes it an interdisciplinary topic of special appeal.

1. Introduction

Symmetry in the visual and geometric sense is omnipresent. It occurs in abundant shapes and forms of nature and hence in the natural sciences. But it also plays a vital role in the arts and makes technology aesthetic. One can also find symmetry in abstract forms, as in quantum mechanics, particle physics, and in relativity theory. Visual symmetry and abstract symmetry both have the same underlying mathematical model, namely the theory of groups. This implies that the notion of groups and subgroups is absolutely essential for understanding symmetry and symmetry breaking.

A fundamental conclusion in the modern understanding of symmetry is the fact, that *symmetry is the rule and not the exception.* Symmetry breaking - as much as symmetry - may manifest itself in a visual and also in an abstract way.

There are two goals in this paper:
- to introduce elementary group theory to students in a stimulating, applied and interdisciplinary setting;
- to convince teachers, that group theory is an important and attractive subject to be treated on the tertiary level, and in parts also on the secondary level.

To achieve these goals in class, I first establish a motivating setting as described in Section 2 and the experiment of the plastic ruler described in Section 6.3. Then the notion of groups is introduced as in Section 5. Here students also work on their own by solving the problems A through F. Finally, the corn circles described in Section 6.2 generate lively discussions, not only during math classes.

2. A Drop of Milk, its Splash and Curie's Principle

Consider a drop of milk falling into a glass of milk. If if hits the milk's surface, its splash looks like a regular 24-indented crown. This can be made visible by a high-speed camera. Students may have encountered such photographs.

The physicist and Nobel prize winner Pierre Curie (1859-1906) realized that many processes are based on principles of symmetry. He stated in 1894 in the *'Journal de physique théorique et appliqué'*:

If certain causes produce certain effects, then the symmetries of the causes reappear in the effects produced.

Studying the drop of milk and its splash, one finds that the cause (the drop) has circular symmetry, while the effect (the crown) does *not* have circular symmetry, but only partial symmetry. Symmetry breaking occurs, proving that Curie was possibly not accurate.

3. Reasons for Symmetry Breaking

The reason for symmetry breaking is the *stability* of the symmetric system. The crown of the splash of milk is stable. But any of the crowns obtained by rotation about an arbitrary angle, can be an effect of the same cause. Rather than one unique effect, symmetric systems have a *set* of possible effects, a fact that is readily understood by students.

Hermann Weyl (1885-1955) gave the following statement:

The truth as we see it today is this: The laws of nature do not determine uniquely the one world that actually exists (Weyl, 1982, page 27).

If so, how does nature choose? It is not *stability* alone, but also the *imperfections* of the symmetry that help nature to decide on its choice. In real life, there are always small *perturbations* that lead to a certain unique effect.

Taking all aspects into account, Stewart and Golubitsky (1992) formulate their *Extended Curie Principle* in chapter 3 in the following form:

Physically realizable states of a symmetric system come in bunches, related to each other by symmetry. Or equivalently: *A symmetric cause produces one form of a symmetrically related set of effects.*

4. Defining Symmetry

It took mankind many thousand years to formulate the concept of symmetry comprehensively. A milestone in history may have been the proof of the existence of precisely 5 regular polyhedra, the so-called Platonian solids. As often in history, names of individuals are associated with results that were found earlier by others. The history of mathematics, as compiled by westerners, on a large scale lacks a thorough analysis, in particular, of what ancient Chinese mathematicians had found long before the Greeks came into play. The notion of *symmetry* is certainly an interesting topic to search for in early Chinese mathematics. In the course of time it became known that symmetry does not only occur in the visual and geometric sense, but also - and in particular - in the *hidden*, more abstract sense of higher dimensional spaces.

The 20th century brought forward the understanding that it was not so much the discovery of symmetry that was new in science, but more so the revelation of its universality.

Experience shows that students are particularly inspired by the topic of symmetry, even those who usually do not like to do mathematics. Some of the reasons are at hand: The phenomenon of symmetry occurs in various fields. It has aspects from geometry, biology, technology, natural sciences, it has aesthetical aspects and hence also from the arts. Symbols as well often show symmetries. It is this abundance that makes for attractiveness to a great number of students more than any other topic, such as arithmetics done with mere numbers or the construction of triangles.

In the classroom, rather than giving the modern, strongly mathematical definition, it makes sense to start with discussing about what the students intuitively understand by symmetry. In the beginning, a possibly confusing number of opinions may have a motivating effect in trying to find an only definition that is accepted by all.

Having found a definition of symmetry by the students in class, it is time to compare it with the abstract, modern definition, which is stated as follows:

A manifold in space and/or time that is invariant under a group G of transformations has the symmetry of the group G.

This makes it obvious that the concept of groups plays a central role in the understanding of symmetry. On the grounds of such a definition, it is clear that abstract symmetry can only be handled by students at the tertiary level with sound knowledge in mathematics. We shall remain elementary and visual in what follows.

The inventor of the concept of groups was Evariste Galois (1811-1832), who died from a duel. His interesting biography is worth a mention in class.

5. Groups and Subgroups

The problems A through F given below at the various stages help the students to obtain an understanding of the concepts. The problems are establihed as having a stimulating and motivating influence in the learning process of the students.

My experience is that the *dihedral group* D_4 is a good example to begin with. It contains all (linear) transformations of the plane that leave an object like the following invariant:

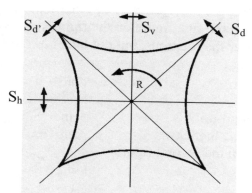

Clearly, $D_4 = \{I, R, R^2, R^3, S_h, S_v, S_d, S_{d'}\}$, where the four S's denote the reflections with respect to straight lines, R is the rotation about 90^0 and I the identity.

In this geometrical example, the students recognize all the ingredients that make up a group, namely that any product (composition) of elements is again an element of the group, that there exists an inverse and an identity.

Problem A: Find the inverse of each of the elements of this group by geometric reasoning.

Problem B: Find the multiplication table of this group by geometric reasoning.

When solving problems A and B, a large number of students is amazed by the fact that not only numbers can be multiplied. There is a remarkable process going on in their minds improving their capability for abstraction.

Problem C: Find the subsets of this group that themselves constitute groups and hence are subgroups.

This problem helps to intuitively deepen the students' understanding of a set being closed under a certain operation.

Problem D: Visualize each subgroup by a figure that has the smallest possible symmetry and that is invariant under the transformations.

There are no restrictions on the students' imagination for finding symmetric figures. A possible answer to C and D could be as follows:

1. The cyclic group $C_4 = \{I, R, R^2, R^3\}$ leaves the structure in the first figure on the right invariant under the rotations, but not under the reflections.

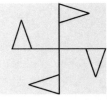

2. The dihedral group
   $D_2 = \{I, R^2, S_h, S_v\}$ leaves the structure in the second figure invariant.

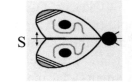

3. The dihedral group
   $D_1 = \{I, S\}$ describes a bilateral symmetry like the one shown in the third figure

4. The cyclic group $C_2 = \{I, R^2\}$ leaves the structure in the fourth figure invariant.

The subgroups of D_4 are thus *visualized* with *less symmetry* than that of the group D_4. This makes it obvious to students that *symmetry indeed is measured by groups*. When asked to compare the various answers to problem D given in class, the students also absorbed the aspect of abstracting the symmetry-grades by the groups that classified their figures.

Problem E: Show that $\{S_v, S_d\}$ is a *generating set* of the group D_4, i.e. each element of the group can be written as a product (composition) of these two elements.

Problem F: For each subgroup give a smallest possible generating set.

Here are two more examples of groups:

- To memorize the spelling of China's capital *Beijing*, it helps to notice, that the middle sequence of letters *iji* is invariant under reflection.

- As legend has it, in ancient China, a divine turtle appeared to the emperor Yu, when he was sitting on the bank of a river, immersed in thought. On the turtle's back, he saw the magic square:

$$4\ 9\ 2$$
$$3\ 5\ 7$$
$$8\ 1\ 6$$

Lines, columns and diagonals are permuted under the transformations of the group D_4. Hence their sums are *invariant*. From this, the emperor Yu obtained another seven magic squares. Among them are

$$2\ 7\ 6 \qquad\qquad 4\ 3\ 8$$
$$9\ 5\ 1 \quad \text{and} \quad 9\ 5\ 1$$
$$4\ 3\ 8 \qquad\qquad 2\ 7\ 6$$

5. Understanding Symmetry Breaking

The concept of symmetry-breaking is, as indicated in the beginning, taken from the natural sciences. It is of fundamental importance for explaining phenomena in nature and technology that are amazing and surprising to the non-experts. Symmetry-breaking uses group theory as an indispensable mathematical tool.

With the concepts of groups and subgroups in mind, the students themselves are now in a position to work out mathematically the phenomenon of symmetry breaking that is summarized in the following diagram:

The cause has the symmetry of a group G	→	The effect has the the symmetry of only a subgroup of G

In other words: The cause has a larger symmetry than the effect. The latter has only a subset of the symmetry of the cause. This is illustrated in the three examples that follow.

6. Examples of Symmetry Breaking

6.1 Once Again: The Drop of Milk

The cause (the falling drop) is invariant under all rotations about the vertical line of dip, f, along the drop's surface as well as under the reflections of the planes through f. The group G is isomorophic with the group O(2) of the orthogonal 2x2-matrices. Sometimes, the notation C_∞ is used for O(2).

The effect most probably, however, has only the symmetry of the dihedral group D_{24}, which leaves a regular 24-gon invariant. We have a dramatic loss of symmetry, from an infinity of symmetry transformations of G down to finitely many symmetry transformations of D_{24}.

Remark: From mathematical models, the number 24 follows indeed. More information on the drop of milk can be found in Thompson (1983).

6.2 Corn Circles

Corn circles were a favorite topic in the Swiss media during the summer of 2000. Bizarre tales with photographs of corn circles were displayed in magazines, newspapers, TV, radio and on the Web. Perhaps not amazingly, mathematics and the concept of symmetry breaking were completely ignored. Similar phenomena were reported also in England. I am convinced that this topic must have been a subject of discussions in other countries, too.

Corn circles are amazingly precise depressions in large and flat fields of standing corn with symmetries that apparently occur spontaneously. Almost each pattern is unique. But they all have the symmetry of a group. Often, it is a dihedral or a cyclic group, sometimes in combination with translations. Symmetry breaking indeed explains the phenomenon to a certain extent. Huge planes (practically considered as an infinitely extended plane) have the symmetry of a large group G of infinitely many elements. The plane is invariant under rotations about arbitrary centers, reflections with respect to any straight line, translations, and combinations of them. G is the group of the proper and improper isometries of the plane. Now, nature has a tremendous choice of subgroups of G. Therefore, we should not be surprised by the abundant variety of corn circles. So, when discussing corn circles in your country, do not ignore the mathematics and substitute it by UFO stories!

6.3 Bending and Bifurcation

For a simple experiment, take a plastic ruler, as used in school. Press on it vertically, as shown in the following figure on the left. If you continuously increase the force F, you notice that at a certain critical force F_c the ruler bends to one side. Which side it is, is a matter of chance. In practice this means the following: Small imperfections of the symmetry will decide to which side it bends. The bending happens because the symmetric case loses its stability, if the force F is greater than the critical force F_c. Theoretically, the symmetric case is always a solution. But it is not stable, hence it is never or rarely observed. This circumstance can be expressed by a *bifurcation diagram*, shown in the figure on the right.

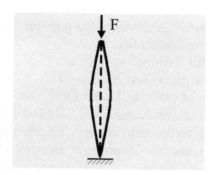
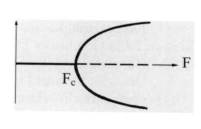

7. More Examples and Informations

In a masterpiece in the popularization of science, Steward and Golubitsky (1992, 1993) discusses many examples of symmetry and symmetry breaking. Here we just mention a few that show the enormous range of applications of group theory, symmetry and symmetry breaking:

- treetrunks may have screw symmetry on the surface;
- in astronomy, elliptical nebulae may bifurcate into different kinds of spirals;
- there are exactly 7 one-dimensional ornamental classes;
- the Taylor-Couette flow;
- Zeeman's catastrophe machine;
- flow around an airplane;
- the Bénard convection in heated liquids;
- symmetry of crystals;

- space-time symmetry (for instance, in a galloping cat or in human walking)

We like to point out to some very interesting links between chaos and symmetry. A nice introduction to this is given by Field and Golubitsky (1992).

Fässler and Stiefel (1992) and Mirman (1995) are more problem-oriented. Weyl (1969, 1982) discusses the notion of symmetry in arts, physics, geometry, algebra and biology in a very nice and appealing way.

References:

Fässler A and Stiefel (1992) 'Group Theoretical Methods and Their Application' Boston: Birkhäuser.

Field M and Golubitsky M (1992) 'Symmetry in Chaos' Oxford: Oxford University Press.

Mirman R (1995) 'Group Theory, an Intuitive Approach' Singapore, New Jersey, Hongkong and London: World Scientific.

Stewart I and Golubitsky M (1992) 'Fearful Symmetry, Is God a Geometer?' Oxford UK and Cambridge USA: Blackwell. Deutsch: Stewart I und Golubitsky M (1993), 'Denkt Gott symmetrisch?' Basel, Boston, Berlin: Birkhäuser.

Thompson D'Arcy W (1972) 'On Growth and Form' Cambridge: Cambridge University Press. Deutsch: Thompson D'Arcy W (1983) 'Über Wachstum und Form' Kurzfassung, neu herausgegeben von John Tyler Bonner, übersetzt von Ella M.Fountain et al mit Geleitwort von Adolf Portmann, Frankfurt: Suhrkamp (Lizenzausgabe Birkhäuser).

Weyl Hermann (1969) 'Symmetry' Princeton: Princeton University Press: first paperback printing from the original (1952) 'Symmetry', Princeton: Princeton University Press. Deutsch: Weyl Hermann (1982) 'Symmetrie' Basel: Birkhäuser.

13

The Rainbow: From Myth to Model

Hans-Wolfgang Henn
University of Dortmund, Germany
wolfgang.henn@math.uni-dortmund.de

Abstract

In all cultures over the world the rainbow is a source for myths and fairy tales. The mathematical theory of the rainbow – in modern language: a mathematical model for the formation of the rainbow (going back to Descartes) – is a very good example for the modelling cycle: from real situation to real model, mathematization, working at the model, the transfer back and forecasts for the underlying problem, validation by "experimentum crucis". Suitable computer software (DGS, CAS) promote this approach already at lower secondary level.

1. The Flood and other myths

The beautiful natural phenomenon of the rainbow has always sparked the imagination of people. Its brilliant bow has always called for interpretation, first synthetically as myths and fairy tales, then analytically through scientific explanations (Boyer 1959). There are many religiously inspired explanations for the rainbow. For the Babylonians, the rainbow was the necklace of the goddes Ischtar. In ancient Greece, the rainbow was seen as the path of the goddess Iris as she traveled as mediator between heaven and earth, the iridescent iris (called rainbow skin in German) still reminds one of this explanation. In many cultures, the rainbow was the path on which the dead went to the next world. In popular belief, there is a treasure hidden at the end of the rainbow. The best known myth can be found in the Old Testament where Jahweh creates the rainbow after the Flood as a sign of reconciliation between God and men. In the Old Testament, Genesis, 9: 16, we read, "And the bow shall be in the clouds, and I shall see it, and shall remember the everlasting covenant, that was made between God and every living soul of all flesh which is upon the earth."

Many painters have created a rainbow. Here it is either a sign of hope or a concomitant of disaster. By the way, the rainbow is very popular with English artists, pointing to the weather and the special relationship of the English with it.

2 The mathematical model

Many scientists of occidental history have tried to give an explanation for the rainbow using the knowledge available at their times. As early as 400 BC Aristotle tried it in his Metereologica. In the 13th century Roger Bacon determined the angle of the rainbow experimentally, but interpreted it incorrectly as reflection. The model of Rene Descartes accepted today stems from the 17th century (Descartes 1637). It uses only reflection and refraction of the light and explains sufficiently the basic structure of the rainbow. We will follow the line of argument of this model of Descartes here. Starting from the real rainbow (Fig. 1) we need to describe the modelling situation more clearly in a real model and will arrive through the process of mathematization at a mathematical model, thereby changing to the abstract world of mathematics. Then we need to deduct mathematical results by mathematical reasoning, which will explain the reality of the observed details of the rainbow.

Fig. 1

A big part of the analysis of the rainbow can be elaborated by the students themselves, for example with the help of worksheets. The following page shows the first of those worksheets from Henn and Wong (1998) (translated in English).

Worksheet 1
Why does a rainbow arise? mathematical model

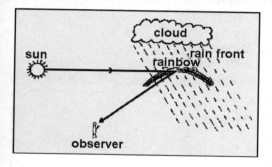

1. What do you know about rainbows? Write down which experiences and knowledge you will probably have to relate to.

2. Explain the sketch containing "sun, observer, rain" (first figure on the right side).

3. Give reasons for the following simplifications in the second figure:

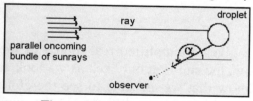

- The original spatial problem can be reduced to a particular plane (which plane?),
- the droplets are spherical,
- the sunrays are parallel,
- the sunrays are scattered by the rain droplets in the direction to the observer,
- the refraction of a sunray can be measured by a suitable angle α.

4. Which angles belong to the "observation area", that means can be seen by the observer?

5. The incident sunrays (on the respective droplet) are partially refracted and partially reflected. Formulate the corresponding law describing the behaviour of a ray which is refracted by a plane (let the angle of incidence be β, the angle of refraction be γ). How can you describe reflection and refraction on spheres?

6. If we suppose that the intensity of a ray is halved when it is reflected and refracted respectively, then you can complete the figure on the right side (the thickness of the lines should symbolise the intensity of the rays). What could be meant by a "ray of class m"?

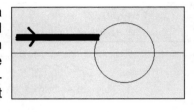

7. The last figure is a more precise version of the second figure above. The scattering angle α depends on the distance x between the incoming ray and the axis of the droplet. x is negative when the ray runs below the axis of the droplet. Unfortunately we don´t know the radius of the droplet. Give reasons why the scattering angle does not only depend on the distance x but also on the ratio x:r. Therefore we can standardize the radius of the droplet and assume r = 1 (why?). Then the so called "impact parameter" has a range $-1 \leq x \leq 1$.

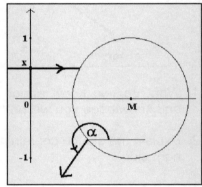

Question 2 of the worksheet gives a simplified real model of the real situation: When we see a rainbow, the sun is at our back as we look at a wall of rain. The light of the rainbow that we are able to see therefore must come from the sun and then somehow change direction at the rainfront towards the observer.

We must idealize this real situation in the mathematical model (questions 3 to 5). We regard sunlight as a parallel oncoming bundle of sunrays whose rays satisfy the reflection law and the refraction law of geometric optics. The rainbow is created by rays, which are scattered in the direction of the observer by the spherical rain droplets. In order to contribute to the rainbow the angle α has to assume values between $180°$ and $270°$. We shall reason in the plane determined by the sun, the centre of the droplet and the eye of the observer. To answer question 6, we distin guish rays of different classes (Fig. 2): The incoming ray is divided into a ray of class one, which is reflected at the surface of the droplet, into a ray of class two which is refracted into the droplet and then immediately refracted out of the droplet, into a ray of class three which is refracted into a droplet, reflected once inside and then refracted again out of the droplet, and so on. For a rough estimate of the intensity we assume that the intensity is divided equally between the reflected and the refracted ray. In the picture this is indicated by an equal thickness of the lines. The intensity of the ray of class n is then only 2^{-n} times the original intensity, which can be neglected soon. Therefore we will consider rays of classes 1 through 4 only.

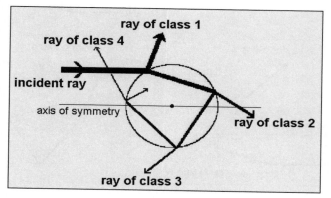

Fig. 2

In the last step of the mathematical model (question 7) we introduce the so-called impact parameter x. Firstly, we standardise the unknown radius of the droplet to 1, which is a meaningful model assumption with respect to the similarity invariance of the situation. A ray of light can only be influenced by a particular droplet when its distance from the axis of the droplet is at most 1. The impact parameter x is therefore a number between -1 and 1. For a given class of the ray the scattering angle α is dependent on x only, therefore we can write $\alpha = f(x)$ with a so-called scattering function f. Now we have constructed our mathematical model, and mathematics can help us to investigate further the four scattering functions.

3. The scattering functions

Because of the symmetry of the situation it is possible to constitute the four scattering functions for the rays of classes 1 through 4 using simple geometric arguments only. Fig. 3 a. – d. show how α depends on the incoming angle β and the refraction angle γ.

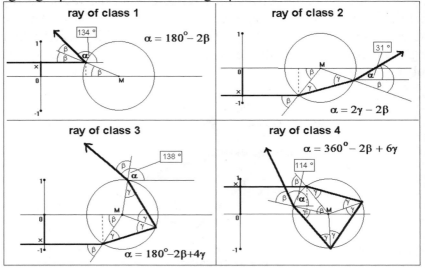

Fig. 3 a. – d.

To investigate the dependence of the angle α from the impact parameter x the way of the rays is constructed using dynamic geometry software (DGS) (we worked with Euklid, which is available via the internet address http://www.dynageo.de) . This is possible at lower secondary level already. We need to find the geometric version of the refraction law experimentally: By measuring the distances d and e instead of the angles in a given circle as shown in Fig. 4, we get $\frac{d}{e} = n = $ constant.

Fig. 4

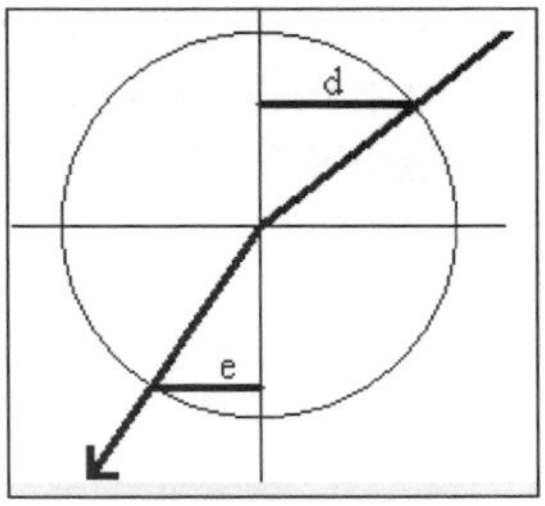

The constant value n is called the refractive index, which has a value of approximately 1.33. Using the DGS we can construct the four paths of light for rays of classes 1 through 4. For the construction of the refraction into the water droplet one needs the geometric law of refraction, all other steps can be done by simple line reflections. Fig. 5 shows the situation of rays of class one.

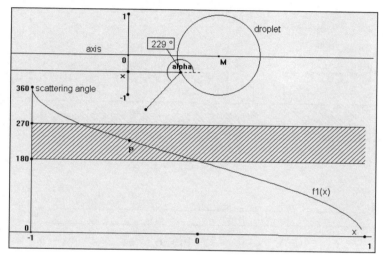

Fig. 5

The angle α is measured by the program; here it is 229°. The basis point x can be dragged to show the dependency of the ray of the light and of the angle α from x dynamically. This dependency is shown in the lower part of the picture as graph of the scattering function. To do this, first the point P is constructed, its x value is the impact parameter x, and its y value corresponds to the angle α. The graph of the first scattering function f_1 emerges as the locus of P, as x assumes values between -1 and 1. The shaded observation area corresponds to widths between 180° and 270°. In this region the graph is nearly linear, light from rays of class one is reflected in all directions with equal intensity. These rays of the highest intensity provide the diffuse background light of the rain front.

Fig. 6 shows the analogous situation for rays of class two, the jump of the graph for x = 0 is caused by the 360° periodicity of the scattering angle α. As one can see there are no rays of class 2 in the shaded region. These rays can therefore not contribute to the rainbow.

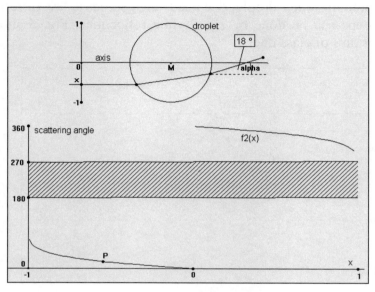

Fig. 6

The construction of the scattering function of class 3 in Fig. 7 explains the genesis of the rainbow. Our rays fall into the shaded observation area from 180° only up to a maximum angle $\alpha^{(3)}$ of about 222°. This angle is a very flat maximum of the curve. Thus, relatively large changes of the impact parameter x hardly change the angle near the maximum. It follows that particularly much light is scattered in an angle area around $\alpha^{(3)}$.

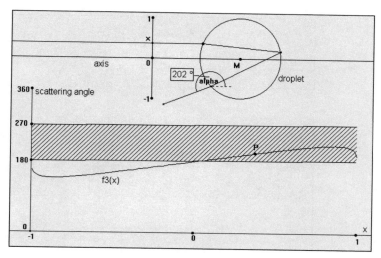

Fig. 7

Fig. 8 illustrates the situation more clearly. We look at three different incoming bundles of rays with equal x widths. The first two are reflected into an angle sector of about the same widths, whereas the third at the extreme value is reflected practically under the same angle $\alpha^{(3)}$.

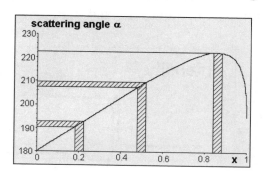

Fig. 8

For rays of class 4 in Fig. 9, we have a similar situation, only that now the graph has a minimum. Our rays fall into the shaded observation area beginning from a minimum angle $\alpha^{(4)}$ of about 230°. This angle is a very flat minimum of the curve.

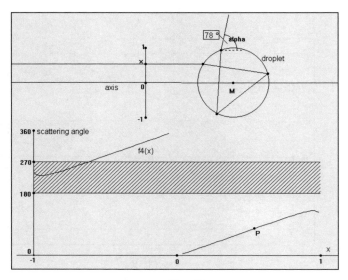

Fig. 9

Together, we have the following distribution of intensities (Fig. 10): The overall intensity I is given by the steady intensity I_1 of rays of class 1 and by the intensities I_3 and I_4 of rays of class 3 and 4, only contributing to I for extreme values $\alpha^{(3)}$ and $\alpha^{(4)}$.

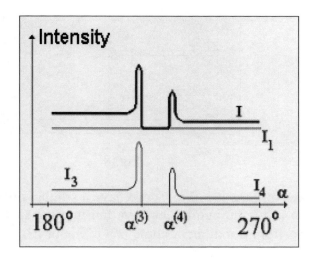

Fig. 10

At these angles we can observe the normally visible primary rainbow and under favourable situations the secondary rainbow. Below the primary rainbow the general intensity is higher, this we observe as a lighter background. In the case that both rainbows are visible one can observe that the sky is darker between them, forming the so-called Alexander's dark band.

The refraction index n and so the angle $\alpha^{(3)}$ are dependent on the colour of the light. Therefore, we see for the primary rainbow red a little higher at the sky than blue, whereas for the secondary rainbow the sequence of the colours is reversed.

So far we have worked in a plane determined by the sun, the centre of the droplet and the eye of the observer. Considering another droplet means that we turn the plane round the axis through the eye of the observer as pointed out in Fig. 11. The point C, under which we can see the rainbow, thereby follows a circular movement, which explains the circular form of the rainbow.

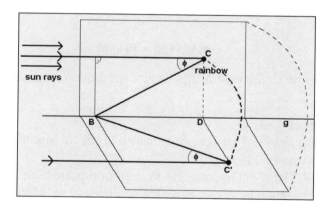

Fig. 11

4. Conclusion

All students should get acquainted with methods and possibilities of mathematics to cope with problems emerging from very different fields of knowledge. The theory of rainbow is such an example of mathematics fraught with relations. The students test the power of the so far developed methods and practise at the same time the transfer

"problem → real model → mathematical model →
mathematical results → solution of the initial problem".

These skills are especially important for the future non-mathematician. Our rainbow model is excellent, every rainbow is a new "experimentum crucis" that is the experiment that confirms the forecasts of the model or not.

Let me close with the shortest and nicest of all country sayings about the rainbow I know, a quotation that suits excellently to ICTMA 10 held in Bejing:

東虹日頭，西虹雨

(A rainbow in the east brings sun, a rainbow in the west brings rain.)

5. References

Boyer C B (1959) 'The Rainbow from Myth to Mathematics' Princeton: University Press

Descartes R (1637) 'Discours de la methode pour bien conduire sa raison et chercher la vérité dans les sciences. Les Météores. De l'arc-en-ciel'

Henn HW and Wong BD (1998) 'Der Regenbogen – ein Projekt im Mathematikunterricht' *Berichte über Mathematik und Unterricht, No. 9807, ETH Zürich*

14

Teaching Inverse Problems in Undergraduate Level Mathematics, Modelling and Applied Mathematics Courses

Fengshan Liu

Department of Mathematics, Delaware State University, Dover, U.S.A..

Email: fliu@dsc.edu

Abstract

The study of inverse ill-posed problems has been a very important factor in the development of applied mathematics, geophysics, technology, medicine and other sciences. However, people need to realize the importance of teaching inverse problems in undergraduate level mathematics, modelling and applied mathematics courses. If students treat all problems as direct problems, they cannot fully understand the problems and therefore cannot solve the problems completely. Using several models from different levels of undergraduate mathematics courses, we demonstrate that the concept of inverse problems could be introduced in mathematics, modelling and applied mathematics courses.

1 Introduction

In this paper, we discuss the importance of teaching inverse problems in undergraduate level mathematics, modelling and applied mathematics courses. Several models from different levels of undergraduate courses will be considered.

In general, a problem may be described by the following diagram:

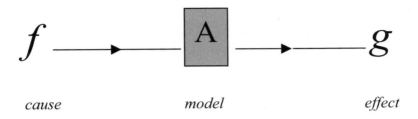

cause   *model*   *effect*

Direct Problems: Given cause f and model A, find effect g;
Inverse Problems: Given model A and effect g, determine f;
Identification Problems: Given cause f and effect g, estimate (identify) A.

When we solve problems, especially Inverse Problems and Identification Problems, we must consider the problem of contamination in data and the effect of approximation. These are not discussed in this paper.

Definition A problem is said to be well-posed if for each effect,
- a solution of the cause exists;
- the solution is unique; and
- the solution depends continuously on the given data (such as effect).

A problem that is not well-posed is called ill-posed. Many inverse problems from mathematical applications are ill-posed. A method for finding stable approximate solutions of an ill-posed problem is called a regularization method. The study of inverse ill-posed problems has been a very important factor in the development of applied mathematics, geophysics, technology, medicine and other sciences. There are several examples: -

1. the derivation of the inverse problem for orbits led to Newton's universal theory of gravitation;
2. mathematical inverse problems are studied in theories of sonar and radar including ground penetrating radar;
3. inverse problems are studied in estimating the earth's girth, weight, interior structure and magnetic field;
4. in medical imaging, the application of inverse Randon's transform is used in the development of computed tomography (CAT)

(see Groetsch, 1993 and Kirsch, 1996) for more examples).

However, people need to realize the importance of teaching inverse problems in undergraduate level mathematics, modelling and applied mathematics courses. We should teach inverse problems in undergraduate level mathematics, applied mathematics and modelling courses.

Actually, we are teaching inverse problems in many mathematics courses, but just have not realized it yet. In this paper, we will study several problems from undergraduate level mathematics courses to illustrate the importance of teaching inverse problems in undergraduate courses. We should introduce the concept of inverse problems and ill-posed problems to our students and teach our students to deal with inverse and ill-posed problems.

By studying inverse problems, students can:
- think "inversely";
- see all sides of a problem;
- deal with ill-posed problems;
- solve problems completely, and
- broaden their intellectual thought.

2. Examples of Inverse Problems

The following examples will demonstrate the importance of teaching inverse problems in the mathematics, modelling and applied mathematics courses.

2.1 Example 1 Measuring the Weight of an Elephant with a Boat

In the ancient times, how did people measure the weight of an elephant, using available technology? (This is an account of a classroom discussion based on an ancient Chinese story in a Business Mathematics class at Delaware State University in spring semester of 1999. Students included freshmen and sophomore.)

Teacher: How would you measure the weight of an elephant using ancient technology? Remember that in ancient times, the traditional hanging bar scale could only weigh maximal of 200 Jin (100 kilograms). When weighing an object, someone needs to lift both the scale and the weighing object up in the air.

Student: An elephant could be a few tons heavy. With the traditional weighing scale, you have to lift the elephant up in the air. You have to have a very long stick to make a huge scale. Well, cut the elephant into pieces, but then you have to kill the elephant.

Teacher: That is right, you do not want kill the elephant. If you want to measure an elephant directly, it would be very difficult because of the elephant's weight. Try to think of a way to measure the weight indirectly. Think of this as an inverse problem.

Student: (Thinking)

Teacher: Boats were available at that time. Different weights on a boat show different water marks on the boat.

Student: Put the elephant on a boat and mark the outside of the boat where water reached as a direct problem, then remove the elephant from the boat and put rocks on the boat until water reaches the mark, weigh the rocks piece by piece. The total weight of the rocks gives the weight of the elephant as an inverse problem.

This example shows that students can solve applied problems indirectly with the concept of inverse problems. Back to the diagram, by treating the level of the boat in the water as g (effect), the weight of the elephant as f (cause), we can set up the model $Af=g$. With the known water level g, we can solve an inverse problem to get the elephant's weight f.

2.2 Example 2 Finding the Number of Pencils

Assume there are two different kinds of pencils k_1 and k_2 with unit costs of c_1 cents and c_2 cents respectively. Denote the total cost of x_1 number of pencils k_1 and x_2 number of pencils k_2 by y. (This is another classroom discussion in a seminar course Applied Mathematics for senior mathematics majors at Delaware State University in the spring semester of 1997. Students who took this course already had Linear Algebra, Discrete Mathematics and Modern Algebra courses.)

Teacher: Let us discuss this problem. Assume that we can only use counting numbers.

Student: The cost function is

$$y = c_1 x_1 + c_2 x_2.$$ (1)

Teacher: You mean you can get the cost if c_1, c_2, x_1 and x_2 are given.

Student: Yes.

Teacher: Ok. Now if you know the unit cost c_1 and c_2, can you find the number of pencils x_1 and x_2 for each integer y?

Student: Yes, of course! Wait, this is an inverse problem, the answer may not be that simple.

Teacher: Let try to look at an example for $c_1 = 8$, $c_2 = 12$, i.e.

$$y = 8x_1 + 12x_2.$$
(2)

Student: Well, y must be even number, (pausing), divisible by 4. There is no solution for $y = 14$. OK, solution may not exist for some integer y for (2). Therefore solution to (1) may not exist for some y in general.

Teacher: Is the solution of (1) unique if the solution exists?

Student: Yes, because the homogeneous case of (1) has only one solution.

Teacher: Good. Now what if you do this problem in the integer number system?

Student: I guess there are more solutions since you can also use negative integers.

Teacher: For this inverse problem, the solution x_1 and x_2 of the equation (1) exist in the integer number system if and only if y is a multiple of the gcd(c_1, c_2).

Here we treat the cause f as a vector of counting numbers (x_1, x_2) and the effect g as the total cost in integer. Then the problem (1) is to solve the inverse problem $Af=g$ for f. This example shows that students should study inverse problems carefully because the solution can be complicated, i.e. solution may not exist or even if it does exist it may not be unique. Furthermore the restriction on the solution also plays an important role. By introducing the concept of inverse problems, we can lead the students to do inverse thinking, to see all sides of the problems, to deal with ill-posed problems and to solve the problems completely. The inverse problem (1) has been studied by Shi X, Liu F, Umoh H and Gibson F (submitted) in the general case of diophantine equations.

2.3 Example 3 Computer Assisted Tomography (CAT)

Computer Assisted Tomography is based on the X-ray principal: as X-rays pass through the body they are absorbed or attenuated (weakened) at differing levels creating a matrix or profile of X-ray beams of different strength. A banana shaped detector measures the X-ray profile. Reconstruction needs to be performed based on the remote measurements from a number of sensors. Such reconstruction requires the solution to an "inverse problem".

To demonstrate the basic mathematical idea of the CAT, let us represent an object by a matrix of "pixels". The problem is to find the elements of the matrix from some input (X-ray) and output (X-ray) information. Assume that $[p_1, p_2 \cdots p_9]$ are the absorption coefficients which are determined by the material of the object.

p_1	p_2	p_3
p_4	p_5	p_6
p_7	p_8	p_9

We send radiation through the object in the direction of $v = [v_1, v_2, \cdots, v_9]$, where v_i is 0 or 1. The fraction p of the radiation that emerges from the object is then $p = (1 - p_1)^{v_1}(1 - p_2)^{v_2} \cdots (1 - p_9)^{v_9}$. We can get nine different equations by sending the radiation in nine different directions. Solving the model $Af=g$ of the system of the nine equations for the cause $f=[p_1, p_2 \cdots p_9]$ with the known effect $g=p$, we can reconstruct the matrix of "pixels". This example shows another real life application of inverse problem theory.

From the above real life examples, we can see that inverse problems are very important in solving modelling problems. In the following, we will show an example of an inverse problem from a mathematics course.

2.4 Example 4 Centroid of a bar

Consider a bar of length L having a nonhomogeneous density distribution. Let the central axis of the bar be the x-axis, the left end of the bar is at the origin and the bar extend to the right. Suppose that the linear mass density of the bar is a given continuous function f. Let $C(x)$ be the position of the centroid of the segment $[0, x]$ of the bar. Then (see Groetsch C (1999) for the detailed modelling procedures),

$$C(x) = \frac{\int_0^x uf(u)du}{\int_0^x f(u)du} . \qquad (3)$$

We treat problem (3) as $Af=g$ where $g=C(x)$, $f=f(u)$ and A is the operator on the right hand side of (3). Finding $C(x)$ with given $f(u)$ in (3) i a simple direct problem. It is clear to see that the solution of the direct problem exists and is unique, that is, given f we can find C. We can also show that if

$$C_n(x) = \frac{\int_0^x u f_n(u)\,du}{\int_0^x f_n(u)\,du}$$

and $\lim_{n\to\infty} \max | f(u) - f_n(u) | = 0$, then $\lim_{n\to\infty} \max | C(u) - C_n(u) | = 0$, where $f_n(u)$ and $C_n(x)$ are successive approximations to $f(u)$ and $C(x)$. This means that the direct problem is stable.

Now, let us consider the inverse problem of finding the density function $f(x)$ with given $C(x)$. As we have discussed in the previous examples, our students will consider the ill-posedness of an inverse problem if we (as teachers) introduce the concept of inverse problems in the courses. Actually the solution of the inverse problem exists, but is not unique. Solutions differ by a constant multiple.

Finally we consider the stability of the inverse problem by considering the following example. Let $C(x)=x/2$ and $f(u)=1$. It is easy to show that $C(x)$ is the centroid function of the density function $f(x)$. Let us model some imperfection in the bar by taking $C_n(x) = \frac{x}{2} + \frac{1}{n} x^{n^2}$ for $n \geq 3$. Let $f_n(u)$ be the density with total mass 1 associated with the centroid $C_n(x)$. We can show that $\lim_{n\to\infty} f_n(1) = \infty$ which implies that $\lim_{n\to\infty} \max | f(u) - f_n(u) | = \infty$. Therefore, the inverse problem is not stable.

Acknowledgement

This research was partially supported by funds from Army Research Office DAAD19-01-1-0738 and Aberdeen Army Research Laboratory DAAD17-01-P-1204.

References

Groetsch C (1999) 'Inverse Problems, Activities for Undergraduates', The Mathematical Association of America.

Groetsch C (1993) 'Inverse Problems in the Mathematical Sciences', Verlag Vieweg, Braunschweig.

Kirsch A (1996) 'An Introduction to the Mathematical theory of Inverse Problems', New York: Springer.

Shi X, Liu F, Umoh H and Gibson F (submitted) *A New Algorithm on Linear Diophantine Equations*.

15

Bezier Curves and Surfaces in the Classroom

Baoswan Dzung Wong
Aargauische Kantonsschule Wettingen, Switzerland
wongdzung@datacomm.ch

Abstract

   Bezier curves and surfaces are the preferred tools of designers in the graphic industry. However, they have been primarily used in automobile, aircraft, ship building, in defining landscapes in geology or road construction, also in plastic surgery and in computer animation. They were independently invented by two engineers of the car building industry, Paul de Faget de Casteljau of Citroen and Pierre Bezier of Renault in the early 1970s. This paper describes some of the features (Sections 1-4) and applications (Sections 5-9) of Bezier curves and surfaces taught to high school students at the Kantonsschule Wettingen, Switzerland. The geometric aspects, (Sections 1, 2) were taught to students in the precalculus phase (aged 16-17), the analytic aspects (Sections 3, 4) involving calculus were taught to older students (aged 18-19).

1. Defining a Bezier curve by geometric construction

   Let the four control points P_0, P_1, P_2, P_3 be given. They generate a curve defined as follows:

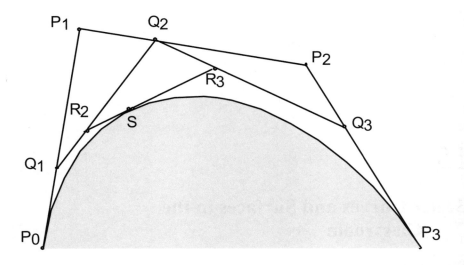

Figure 1: Construction of the point S of the Bezier curve from P_0, P_1, P_2, P_3

The line segments $\overline{P_0 P_1}$, $\overline{P_1 P_2}$, $\overline{P_2 P_3}$ are each divided at a given ratio $t : (1-t)$, where $0 \le t \le 1$, to obtain the points Q_1, Q_2, Q_3 (see Figure. 1). These points in turn are joined to form segments which are divided at the same ratio $t : (1-t)$ to yield R_2, R_3. Applying this procedure to $\overline{R_2 R_3}$ results in a single point S. If t varies within the range [0,1], the corresponding points $S(t)$ trace a curve, called the *Bezier curve of third order*. The Bezier curve passes through the points P_0 and P_3 and is tangent to $P_0 P_1$ and $P_2 P_3$.

This algorithm applied to $n+1$ control points P_0, ..., P_n yields a *Bezier curve of n-th order* and was first conceived by Paul de Faget de Casteljau. In the graphics industry, it is enough to work with Bezier curves of third order (i.e. with four control points, as shown in Fig. 1), in the moulding industry one may need Bezier curves and surfaces of up to the fifth order.

2. The parameter representation of a Bezier curve

The equation of a Bezier curve of third order is obtained from de Casteljau's algorithm:

$$Q_i = (1-t) \cdot P_{i-1} + t \cdot P_i \qquad \text{where } i = 1, 2, 3$$
$$R_i = (1-t) \cdot Q_{i-1} + t \cdot Q_i \qquad \text{where } i = 2, 3$$

are substituted into the equation $S = (1-t)\cdot R_2 + t \cdot R_3$ to obtain a vector-valued function depending on the variable t:

$$S(t) = \sum_{i=0}^{3} \binom{3}{i} \cdot (1-t)^{3-i} \cdot t^i \cdot P_i \quad \text{where } 0 \le t \le 1 \tag{1}$$

This is the *parametric representation* of a third-order Bezier curve.

$S(t)$ can be interpreted as the weighted average of the control points P_0, P_1, P_2, P_3. The weight functions

$$B_i(t) = \binom{3}{i} \cdot (1-t)^{3-i} \cdot t^i$$

are the third order *Bernstein polynomials* known from the Weierstrass approximation theorem. They have a number of favourable properties that explain why Bezier curves are preferred over polynomial interpolations:

(i)   Bezier curves always lie within the convex hull of their control points. This guarantees that the curve does not wiggle around the given control points, as may happen with polynomial interpolation.

(ii)   Bezier curves do not depend on the underlying coordinate system as do the polynomials obtained from interpolation.

3. Composing Bezier segments

With the parametric representation (1), one can compute its derivatives. These are needed when assembling two Bezier segments to form a composite curve. If smooth transition at the shared endpoints is desired, then the following conditions must hold:

(i)   Evidently, the transition from the curve $S(t)$ to the curve $T(t)$ is continuous, if the last control point P_3 of $S(t)$ coincides with the first control point Q_0 of $T(t)$.

(ii)   The first derivative is continuous, if the point of transition $P_3 = Q_0$ is the midpoint of the adjacent control points P_2 and Q_1.

(iii)   The second derivative is continuous, if the third to the last control point P_1 of $S(t)$ and the third control point Q_2 of $T(t)$ are such, that the vector $\overline{P_1 Q_2}$ is parallel to and twice as long as the vector $\overline{P_2 Q_1}$.

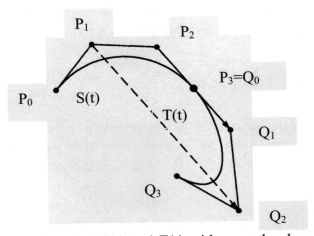

Figure. 2: Two Bezier segments $S(t)$ and $T(t)$ with second order smoothness

In general, if a Bezier curve has order of at least $k+1$ and continuity of order k is desired, one has to observe conditions that involve the last $k+1$ control points of $P(t)$ and the first $k+1$ control points of $Q(t)$.

4. Bezier surfaces

A Bezier surface can be regarded as a surface consisting of curves with control points that move along Bezier curves, see Fig. 3.

Each of the sets P_{0j}, P_{1j}, P_{2j}, P_{3j}, for $j=0,...,3$, generates a Bezier curve represented by

$$P_j(t) = \sum_{i=0}^{3} B_i(t) \cdot P_{ij} \quad \text{where} \quad j=0,1,2,3 \text{ and } t \in [0,1]$$

The points $P_0(t_0)$, $P_1(t_0)$, $P_2(t_0)$, $P_3(t_0)$, for some $t_0 \in [0,1]$ are shown in Fig. 3. They, in turn, are considered as control points, that generate a Bezier curve

$$Q(s,t_0) = \sum_{j=0}^{3} B_j(s) \cdot P_j(t_0) \quad \text{where} \quad s \in [0,1]$$

The surface can be described by a vector-valued function depending on two variables:

$$Q(s,t) = \sum_{j=0}^{3}\sum_{i=0}^{3} B_j(s)B_i(t)P_{ji} \quad \text{where} \quad s,t \in [0,1] \qquad (2)$$

This is the parametric representation of a third-order *Bezier surface*.

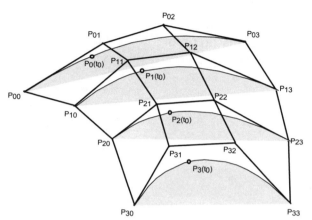

Figure. 3: The control points of a Bezier patch

5. Embroidering de Casteljau's algorithm

Some students gather their first experience with the geometry of de Casteljau's method by drawing or embroidering it, as shown in Fig. 4.

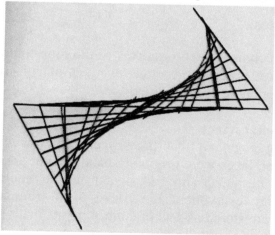

Figure 4: De Casteljau's algorithm embroidered

The initial line segments $\overline{P_0P_1}$, $\overline{P_1P_2}$, $\overline{P_2P_3}$ are divided evenly into ten segments and the dividing points labelled successively 1, 2, ..., 10.

Joining the points carrying the same number yields the intermediary segments or threads. This operation is repeated upon the intermediary lines to yield a second set of construction lines or threads.

6. Programming a Bezier curve

Since a large number of de Casteljau's constructions are needed to produce a Bezier curve, it is favorable to let the students write a program on their pocket calculators, e.g.

```
PROGRAM:CASTLJAU
:ClrLCD
:ClDrw
:ZSq

:Input "P0X=", P0X
:Input "P0Y=", P0Y
:Input "P1X=", P1X
:Input "P1Y=", P1Y
:Input "P2X=", P2X
:Input "P2Y=", P2Y
:Input "P3X=", P3X
:Input "P3Y=", P3Y

:For (T,0,1,0.05)
:(1-T) → L
:L* P0X +T* P1X → Q1X
:L* P1X +T* P2X → Q2X
:L* P2X +T* P3X → Q3X
```

```
:L* Q1X +T* Q2X → R2X
:L* Q2X +T* Q3X → R3X
:L* R2X +T* R3X → S3X
:L* Q1Y +T* Q2Y → R2Y
:L* Q2Y +T* Q3Y → R3Y
:L* P0Y +T* P1Y → Q1Y
:L* P1Y +T* P2Y → Q2Y
:L* P2Y +T* P3Y → Q3Y
:L* R2Y +T* R3Y → S3Y
:Line(P0X, P0Y, P1X, P1Y)

:Line(P1X, P1Y, P2X, P2Y)
:Line(P2X, P2Y, P3X, P3Y)
:Line(Q1X, Q1Y, Q2X, Q2Y)
:Line(Q2X, Q2Y, Q3X, Q3Y)
:Line(R2X, R2Y, R3X, R3Y)
:PtOn(S3X, S3Y)
:End
```

Upon receiving the co-ordinates of the four control points, P_0, P_1, P_2, P_3, the above program generates a mesh of construction lines that wraps in a Bezier curve of third order.

7. Letters and characters

Lettering is an art that takes years for a well-trained artist to develop. All type faces were originally hand drawn, engraved in stone or cast in metal, designed and redesigned in the endeavor to achieve perfection. Today everyone can retrieve them from a computer. The computer fonts are stored as sets of control points that generate the letters by means of Bezier curves. In the classroom, my students created on their pocket calculators the letter Q, both in normal and in slanted type style. The contours of the letter Q were composed of six second or third order Bezier segments. The slanted type style was obtained from the normal one by shearing the six control polygons with respect to the horizontal axis, see Fig. 5.

Figure 5: The letter Q generated by Bezier segments and their control polygons.

Finally, an example of a Chinese character rendered by Bezier curves. Here all of the Bezier segments used are of third order.

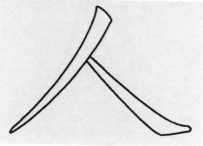

Figure 6: The Chinese character "ren" (human being)

3. An application of Bezier surfaces

Aladdin, in Walt Disney's motion picture, has a number of good friends who help him overcome the hardships in his life. Among them is a carpet, though without limbs or a face, it exquisitely expresses its feelings by the mere bending of its body, the two-dimensional carpet itself. As a first example, the students try to model Aladdin's carpet by a single Bezier patch, e.g.

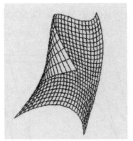

Figure 7: Aladdin's carpet

9. An oil bottle

The horizontal sections of the oil bottle shown in Fig. 8 are circle that were approximated by Bezier segments. The vertical sections along the bottle's axis were drawn on a piece of paper, then measured and converted into Bezier segments. The bottle consists of 16 Bezier patches Their parametric representation, as given in equation (2), is rendered by use of the MAPLE Computation System.

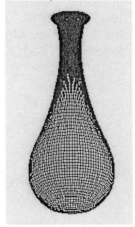

Figure. 8: An oil bottle modelled by use of Bezier patches

Acknowledgement.

This study was made possible by a joint grant of the "ETH fuer di Schule" and the Department of Education of the canton Aargau Switzerland. I wish to thank Prof. Dr. Urs Kirchgraber, ETH Zurich, fo supervising my work. I am also obliged to Dr. Albert Faessler, HTA Biel/Bienne, who carefully read my manuscript.

Section E

Teacher Education

16

A Mathematical Modelling Course for Preservice Secondary School Mathematics Teachers

Zhonghong Jiang, Edwin McClintock and George O'Brien
Florida International University
Jiangz@fiu.edu

Abstract

This article describes a mathematical modelling course designed for preservice mathematics teachers. The purpose of this course is to provide the future teachers with the knowledge and experience that enable, motivate, and encourage them to solve real-world problems. The course content is aligned with the National Council of Teachers of Mathematics Standards Documents (1989, 2000). To stimulate the students' interest in mathematics and develop their spirit of active thinking and inquiry, the course features problem solving (especially mathematical modelling), mathematical inquiry and extensive use of technology. The curriculum approach is providing opportunities for the preservice teachers to experiment with abstract concepts, objects, and relationships, and pursue conceptual understandings.

1. Introduction

College level mathematics courses contain theory and a certain degree of rigour. Many of our preservice teachers, who are weak in mathematical reasoning, find it hard to successfully pass courses such as College Geometry and Number Theory. While theory and rigour are necessary, our own way of teaching and our grading practices drive these students away from the pursuit of mathematics at an excessive rate. Most courses taught by the mathematics faculty are very traditional, lecture-oriented courses, which, without change, could not help our preservice teachers to improve themselves academically.

It is time to reverse the questionable performance in the mathematics preparation of our preservice teachers. To provide them with quality mathematics education, we have focused on curriculum changes. One of the important changes is designing and implementing a new, standards-based, mathematical modelling course for secondary school mathematics preservice teachers. This course can help the prospective teachers construct their content knowledge from a perspective that involves rich connections among mathematics, science, and real-world situations. The course features innovations derived from the national mathematics education standards. It has been and will continue to be offered at the junior level, and is team-taught by a group of faculty using exemplary teaching strategies. The knowledge, skills, and thinking methodologies learned in this new course will be carried through the development of other new courses or course components that should impact the future teachers in the following years.

2. Course Description

The purpose of this course is to provide future teachers with the knowledge and experience that enable, motivate, and encourage them to solve real-world problems through mathematical modelling.

2.1. Course guideline

The course content is aligned with the National Council of Teachers of Mathematics (NCTM) Standards Documents (1989, 2000). One of the guiding ideas of this course is mathematical modelling as conceptualized by the NCTM "Mathematics as Problem Solving" Standard in the following diagram (1989, p.138):

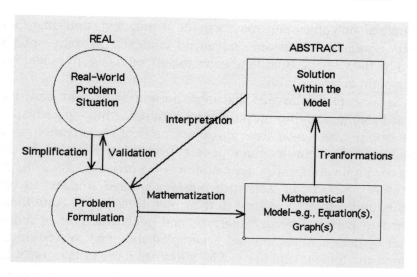

The course is also based on current research and theories about how students learn mathematics, which include the constructivist learning theory and the research on mathematical modelling and students' motivation. Learning as a constructive process suggests new learning is an interaction among what an individual knows, what information is encountered, and what individuals do as they learn. Rather than passive recipients of knowledge and skills (the result of lecture and explanation), students learn best when they try to make sense of new information and relate it to what they already know. Unless new information is linked to and integrated with the learner's prior knowledge and understanding, this new information remains isolated, cannot be used in addressing new tasks, and does not readily transfer to new problems or situations (Lambert and McCombs, 1998).

Calls for restructuring learning environments that build around genuine mathematical inquiry has increased substantially over the past few years. Explicit in the current recommendations for reform in school mathematics is the notion of helping learners develop the disposition to engage in activities similar to those of mathematicians (NCTM, 2000). Current convictions shared by many mathematicians and mathematics educators about the nature of mathematical knowledge and how it is best learned include, inquiry into real life questions, planning problem and data driven curriculum that motivates the search for meaning and the use of mathematics (National Research Council, 1995). Using modelling as an instructional and learning strategy is a promising tool for improving mathematical outcomes and may provide a practical, unifying approach that will create the necessary "advanced context". A study soon to be released by the College Board indicates the effectiveness of modelling as a teaching approach.

Sarason (1996) argues that motivating a student to learn may be the most basic underlying purpose for education. More specifically, it is the classroom experience that impacts a student's motivation to learn. McCombs (1998) identifies three general principles related to motivation that have implications for the design of curricula as well as for instructional practice: 1) Activities must encourage students to become personally and actively involved in their own learning; 2) Activities must be related to personal needs, interests, and goals (directly or indirectly) and must be both challenging and accomplishable; and 3) Activities must be tailored to students' unique needs, allow students to take risks and be facilitated by a caring adult who sees the students' potential for success.

We no longer can rest on a past in which memorization of facts and figures dominated the learning environment. Today, students, especially our preservice teachers, must be able to solve problems, make decisions, work with their peers, and pursue new learning as ideas evolve. Mathematical modelling activities and building the motivation to learning can influence students' interest in learning for a lifetime.

2.2 Course features

To stimulate students' interest in mathematics and develop their spirit of active thinking and inquiry, the course features problem solving (especially mathematical modelling), mathematical inquiry and extensive use of technology. The curriculum approach is providing opportunities for the preservice teachers to experiment with abstract concepts, objects, and relationships, and pursue conceptual understandings.

Mathematical modelling is the main feature of the course. Modelling refers to making sense of data. A model is a simplified representation of a phenomenon that suggests how the phenomenon works. There are three types of models: physical, conceptual, and mathematical. Physical models are actual devices or processes that behave like phenomena. Conceptual models link the unfamiliar by using metaphors and analogies. Mathematical models specify a relationship between variables described and the behavior of phenomena. Mathematical models are more abstract than physical and conceptual models (Krajcik et al 1999). Building models is the central activity. In building a model, one must engage in critical thinking. The language of mathematics such as algebra and differential equations is usually used to express the models. Since modelling is so important, students enrolled in this course experience substantial model exploration and model building (mainly mathematical model building) activities. They are given the opportunity to evaluate and interpret data (activities that precede model building), build models, solve models (functions, graphs, systems of equations, etc.) and communicate findings. Computer tools will be used, which allow students to develop an understanding of mathematical concepts, especially when the students are lacking mathematical sophistication.

The course is inquiry based featuring a series of curriculum units (each 1-2 weeks of activities) for which students complete projects. In each of the units, the content and activities in mathematics, science, and technology are integrated and there is a well represented progression of ideas and skills. Each unit includes objectives, experimental learning, appropriate use of technology and evaluation instruments. Interdisciplinary ideas are required for completing the projects. Web-based components of each unit will help student teams to emerge and student-instructor dialogues to shape a community of active learners. The materials and activities are delivered through a hands-on and minds-on approach, and motivate students to learn the foundational mathematics concepts by involving them in enriched learning experiences relevant to their daily lives. Students identify questions that can be answered with their modelling activities. While emphasizing the learning of concepts, problem solving and mathematical investigation, this course also incorporates best teaching practices.

The topics for the curriculum units will include airline or car routes, traffic patterns, population predictions, and optimization problems (e.g., street-parking design, storage at minimum expense), etc. Mathematics content involved will be general undergraduate mathematics, precalculus (a variety of functions), calculus (derivative and integration), differential equations, data collection and analysis, curve fitting/regression, and so on. Course activities will follow a sequence starting with simple modelling situations. The level of sophistication and difficulty of problem contexts will increase gradually.

Group discussions and class presentations will be the major components of the instruction, which allow students to exchange their ideas of how to understand the problem situations, compare their strategies, and ultimately question each other's solutions and models.

2.3. A sample mathematical modelling activity
    The following is a sample activity designed for a course unit (Blumsack, 2002):
    "Task: Devise an experiment to determine the radius of the earth.
    Discuss how to complete this task in pairs or small groups.
    If you can come up with an idea to devise such an experiment, start working on it and compare it to the one that follows. If you need help, continue reading.
    Suppose an individual in city A erects a pole two meters long vertically and measures the minimum shadow length during the middle of the day to be 50 cm. A colleague in city B located 150 km north of city A repeats the experiment and obtains a measurement of 55 cm.
    The measurement at city A implies that the sun is overhead at a point (called the subsolar point) at a latitude arctan(50/200)=.245 radians south of city A. The corresponding measurement at city B means that city B is arctan(55/200)=.268 radians north of the subsolar point.
    The diagram below illustrates the situation at either of the two cities. The sun is far to the right of the figure with its rays striking the earth horizontally. Notice the relation between the latitude difference and one of the angles in the right triangle associated with the pole and its shadow.

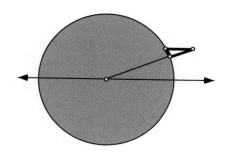

Show that use of these results would imply that the radius of the earth is about 6520 km."

2.4. An example of students' work

The students enrolled in this course were asked to build a mathematical model for the world population growth based on the data from the web site www.census.gov/ipc/www/worldhis.html, which gives historic estimates of world population from 10000 BC to 2000 AD. To make better sense of the population growth, the students used Physics Analysis Workstation (PAW), an interactive graphing and curve fitting system (http://paw.web.cern.ch/paw/), to construct the graphic representations of the data, shown in Figures 1 and 2.

In the graph displayed in Figure 1, the part with the darker marks at left shows the world population from 10000 BC to 1 AD, the part with the lighter marks shows the population from 1 AD to 1950 AD, and the part with the darker marks at right shows the population from 1950 AD to 2000 AD. From the graph, the students clearly visualized that the population of the world had changed slowly until around 1800 AD, and it rapidly grew during the last 50 years.

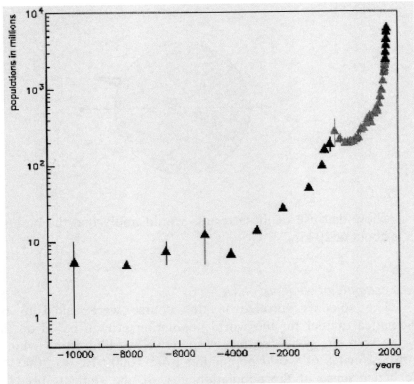

Figure 1. Population growth from 10000 BC to 2000 A.D.

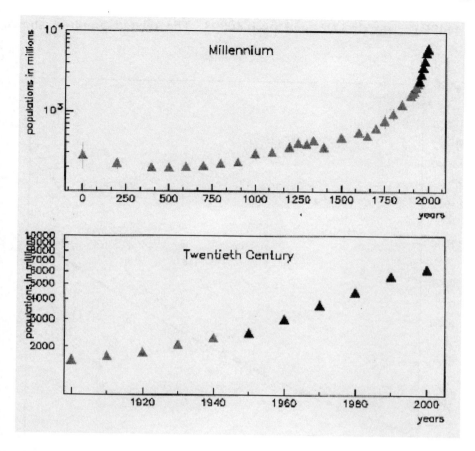

Figure 2. Population growth during the period from 1 to 2000 AD and the 20th century

    In the two graphs displayed in Figure 2, the students could visualize the population change during the period from 1 AD to 2000 AD and that during the twentieth century.

    Analysis of data usually involves fitting the measured data to a model in order to make some predictions about the system under investigation. The students decided to use an exponential function and a quadratic function to model the data. In PAW the fitting algorithm solves the related equations to determine the best-fit parameters for the chosen functions based on some data and a linear model. With the help of PAW, the students got to know that for the exponential function $\exp(p_1 + p_2t)$, the best-fit parameters were $p_1 = -29.263$ and $p_2 = 0.1902\text{E-}1$; and for the quadratic function $p_0 + p_1t + p_2t^2$, the best-fit parameters were $p_0 =$

0.14449E+7, p_1=-1537.2, and p_2=0.40893. The related graph is shown below:

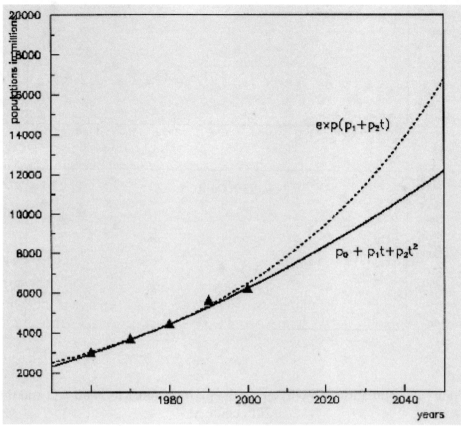

Figure 3. Curve fitting and extrapolation

With the exponential and quadratic functions mentioned above, the students were able to predict the population of the world in the years 2010, 2020, 2030 and 2050. Using the Mathematica software for numerical calculations, the students' predictions are listed in the following table:

	Year 2010	Year 2020	Year 2030	Year 2050
Calculated with the exponential function.	t=2010 Pe = 7841.61* 10^6 Population: around 7.84 billion	t=2020 Pe = 9484.37* 10^6 Population: around 9.48 billion	t=2030 Pe = 11471.3* 10^6 Population: around 11.47 billion	t=2050 Pe = 16781* 10^6 Population: around 16.78 billion
Calculated with the quadratic function.	Pq = 7246.09* 10^6 Population: around 7.25 billion	Pq = 8353.97* 10^6 Population: around 8.35 billion	Pq = 9543.64* 10^6 Population: around 9.54 billion	Pq = 12168.3* 10^6 Population: around 12.17 billion

From these calculations and also from the graph shown in Figure 3, the students realized that during the period 1950-2000 these two functions give close results, but as the value of t increases, the difference between Pe and Pq grows. Thus they understood that when making predictions, one must be careful. To them, it seemed to be a good idea to predict the number of people living on earth within the range of Pe and Pq values.

2.5. Technology integration

The use of technology is emphasized in this course. Calculator-based Laboratory (CBL) with multiple sets of probe-ware is used for data collection. Computer software (such as Mathematica and statistics software) and graphing calculators are used for data analyses and curve fitting activities. Other computer applications such as the Geometer's Sketchpad (GSP) and Microsoft Excel spreadsheet are also used to stimulate the future teachers' mathematical modelling and reasoning insights and their learning interest. These technology tools enable the future teachers to appreciate the power of mathematics that helps them understand the world. Using the graphical and numerical representations together, the future teachers can interpret situations both visually and numerically. This helps them formulate and refine problems (if the problems do not arrive neatly packaged), investigate problems from multiple perspectives to gain further insights, and articulate problems clearly enough to build mathematical models. When they experience difficulties in the problem posing and solving processes, constructed computer situations can help them develop ideas and strategies to approach solutions. These computer situations are usually difficult for the future teachers who lack a sound understanding of the problems to construct a model by themselves in the first place.

The students' modelling activity on the population growth (see above) gives an example of using data plotting and curve fitting software in presenting and implementing modelling ideas. The example below shows how the use of GSP (a dynamic geometry software package) helps students develop mathematical modelling insights.

The STREET PARKING problem:

> *You are on the planning commission for Algebraville, and plans are being made for the downtown shopping district revitalization. The streets are 60 feet wide, and an allowance must be made for both on-street parking and two-way traffic. Fifteen feet of roadway is needed for each lane of traffic. Parking spaces are to be 16 feet long and 10 feet wide, including the lines. You job is to determine which method of parking – parallel or angle – will allow the most room for the parking of cars and still allow a two-way traffic flow. (You may design parking for one city block (0.1 mile) and use that design for the entire shopping district.)*

When exploring this problem, a considerable number of students (our preservice teachers, the same hereafter) intuitively conjectured that angle parking is better than parallel parking (because they mostly experience angle parking). The parallel parking situation is quite simple, and students were able to quickly determine how many cars can be parked in one city block. However, most of the students had difficulty formulating a mathematical solution for the angle-parking situation, other than the intuitive conjecture. We presented a constructed GSP sketch (shown below) for students to investigate the situation.

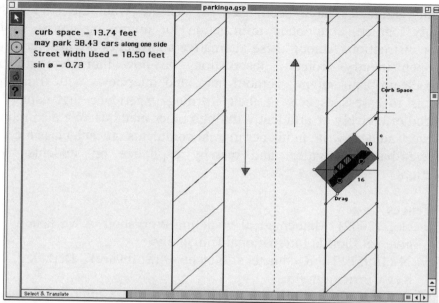

Figure 4. A GSP sketch to help students understand the angle-parking situation

By working with the sketch, the students clearly saw that in order not to block the traffic, the left side of each parking space has to be "dragged down" so that the shaded rectangle does not intersect the lane of traffic (or at most "touches" the traffic lane at one point only). Their action caused the curb space to be very long. The result was that much more space was wasted and fewer vehicles could be parked in one city block. This visual feedback helped the students correct their misconceptions. Moreover, their experience with the situation enhanced the change that they would be able to identify key factors in the situation. Using similar-triangle relationships and the Pythagorean Theorem, students finally arrived at a mathematical solution, and understood why parallel parking is often used in downtown areas where streets may be narrow.

3. Conclusion

We have found that this course can effectively help our preservice teachers reach a better understanding of mathematical concepts and develop stronger problem-solving abilities. To show the evidence, course assessment is very important. In addition to traditional assessment techniques such as quizzes, exams, library research reports, and activity write-ups, alternative assessments based upon numerous alternative learning strategies have been used to measure what the student has actually learned, and obtain information for improving course design and/or instruction. Among these alternative approaches are artifacts from the open inquiry projects, observations on how individual students approach problem solving without help, and interviews with individual students to assess aspects of their learning experience that cannot be revealed effectively or efficiently through other methods. We also require that every student write in his/her journal comments on each class meeting and web-based activities, and thereby capitalize on students' self-reflection.

References

Blumsack S (2002) 'Mathematical Modeling Workshop' A workshop given at Florida International University.

Jackiw N (1995) 'The Geometer's Sketchpad (Software)', Berkeley, CA: Key Curriculum Press.

Krajcik J, Czerniak C and Berger C (1999) 'Teaching Children Science: A Project-based Approach', Boston: McGraw-Hill College.

Lambert N and McCombs B (1998) 'How Students Learn: Reforming Schools through Learner-Centered Education', Washington, DC: American Psychological Association.

McCombs BL (1998) 'Integrating Metacognition, Affect, and Motivation in Improving Teacher education' in McCombs BL and Lambert N (Eds) *How Students Learn: Reforming Schools through Learner-Centered Education*, Washington, DC: American Psychological Association.

National Council of Teachers of Mathematics (1989) 'Curriculum and Evaluation Standards for School Mathematics', Reston, VA: The Council.

National Council of Teachers of Mathematics (2000) 'Principles and Standards for School Mathematics', Reston, VA: The Council.

National Research Council (1995) 'Allocating Federal Funds for Science and Technology', Washington, DC: The Council.

Sarason S (1996) 'Revisiting "The Culture of the School and the Problem of Change"', New York: Teachers College Press.

17

Mathematical Modeling in Teacher Education

Mikael Holmquist and Thomas Lingefjärd
Göteborg University, Sweden.
Mikael.Holmquist@ped.gu.se and Thomas.Lingefjard@ped.gu.se

Abstract

There are many ways to view the teaching and learning of mathematical modeling. In teacher education, the need to know mathematical modeling arises in part from the fact that mathematical modeling is mentioned in many secondary curricula around the world. Today there is an obvious need to understand how to teach and assess mathematics in a technology-enriched environment. At Göteborg University, prospective teachers took part in a course in which the mathematical content was designed to give them insight into how they could solve extended problems using mathematical modeling by drawing on technology and their background in mathematics. Different software and graphing calculators were used. A fundamental idea in the assessment of the students was that assessment and teaching were integrated.

Because of our experiences in earlier studies of the teaching and learning of mathematical modeling (Lingefjärd, 2000; Lingefjärd and Holmquist, 2001), we decided to study whether the type of modeling process that students are involved in causes or at least plays a major role in the conceptions they develop.

The formulation and presentation of mathematical modeling situations, the prospective teachers' views of mathematical modeling, their conceptions of the mathematics involved in the modeling process, including the possibility of validating a complicated mathematical model, are all discussed in this paper.

1. Introduction

The Swedish school should, in its teaching of mathematics, strive to see that students' projects and group discussions develop their conceptual capacity and that they learn how to formulate and argue for different methods for solving mathematical problems.

They should also develop their aptitude to shape, refine, and use mathematical models together with a critical estimation of the model's conditions, possibilities, and limitations. (Skolverket, 2000, p. 2, our translation).

A national statement like the one above raises many questions, including how to prepare prospective secondary school mathematics teachers to function in an environment where teaching and learning are characterized by processes and activities. This statement also reflects new trends in how to look at teaching and learning in mathematics. From a researcher's view it is important to try to clarify and describe different approaches to teaching and learning in mathematics. For example, Sfard (1997) emphasizes that our thinking about learning may be rooted simultaneously in two different conceptual domains:

On the one hand, there is the view of learning as an acquisition of some private property; and on the other hand there is the idea of learning as becoming a participant in a certain practice or discourse. (p. 120)

In this paper we adopt the latter view. The curriculum and syllabus for mathematics in secondary school as well as for teacher education in Sweden emphasizes this "shift" in views of learning. Terms like *knowledge* have been replaced with the noun *knowing* coupled with the verb *use*, thereby clearly indicating action.

So far, there is no agreement in the community of mathematics and mathematics education on what mathematical thinking really is. In *The Nature of Mathematical Thinking* (Sternberg, 1996), Sternberg wrote in the culminating chapter, "In reading through the chapters of this volume, i becomes clear that there is no consensus on what mathematical thinking is nor even on the abilities or predispositions that underlie it." (p. 303) Schoenfeld (1992) used the word *mathematization* in his endeavor to describe mathematical thinking:

Learning to think mathematically means (a) developing a mathematical point of view -- valuing the processes of mathematization and abstraction and having the predilection to apply them, and (b developing competence with the tools of the trade, and using those tools in the service of the goal of understanding structure -- mathematical sense-making. (p. 335)

The process of mathematization might be seen as strongly connected to the situation in which we are applying mathematics in a real-world problem. Working on such a problem means that we are dealing with a collection of objects, relations between these objects, and structures belonging to the area that we are studying. Next we have to translate these into mathematical objects, relations, and structures as representations for the original ones. This entire process, mathematical modeling, leading from the real problem situation to the mathematical model, might be seen as mathematization. But de Lange (1996) gives the term *mathematization* a more restricted meaning by defining it as the translation part of the modeling process. When transferring a problem, especially an investigational problem, to a mathematically stated problem, one engages in some important activities. These might include formulating and visualizing the problem in different ways, discovering relations, and transferring the investigational problem to a known mathematical model. The next step is to attack and treat the mathematical problem with mathematical tools in activities like representing a relation in a formula, refining and adjusting models, combining and integrating models, and generalizing. The students' conceptions of the mathematics involved in the modeling process are essential and are revised through their activities and their reflections on their actions.

Mathematization always goes together with reflection. This reflection must take place in all phases of mathematization. The students must reflect on their personal processes of mathematization, discuss their activities with other students, must evaluate the products of their mathematization, and interpret the result. (de Lange, 1996, p. 69)

There are even authors who claim that instruction that strongly emphasizes structured drill and practice on discrete, factual knowledge can do students a major disservice (Marton and Booth, 1997; Ramsden, 1992). Let us modify this claim by saying that mere acquisition of knowledge skills is not sufficient to make one into a competent thinker or problem solver. Students also need to acquire the disposition to use their skills and strategies, as well as the knowledge of when and how to apply them. These are also appropriate targets for assessment.

2. Assessment

To assess mathematical modeling is not easily accomplished. The more complicated and open the problem is, the more complicated it is to assess its solution by conventional means.

We are prepared to risk our skin by claiming that assessment of applications and modelling is easy. As mentioned earlier, assessment is not easy if we (have to) stick to conventional modes and practices. In that case sound assessment is rather very difficult if not impossible. (Niss, 1993, p. 48)

If one adds the component of existing technology, assessment becomes even more complicated. The support to be provided by technology when students are being assessed is a difficult issue and is the subject of ongoing discussion in several places around the world. A phrase often mentioned together with the use of technology is *authentic assessment* or *authentic performance assessment,* which, according to Clarke (1996), refers to mathematical tasks that are meaningful for the student, represent applications of mathematics, and include activities that are, in some sense, also carried out by mathematicians. We have to find ways of assessing what is looked at as important, rather than assessing what is easily measurable.

The involvement of the students in the assessment is likely to shape the educational process. When students become more involved in the process of evaluation, it may be seen as a substantial part of the didactical contract being negotiated between student and teacher. Through this interplay, the students can learn to identify the criteria for qualitatively good performance. Further, they can also learn the criteria for unsatisfactory, fair, good, or very good performance and how to differentiate between these criteria. It makes sense to give learners opportunities to analyze strong and weak answers to more open-ended problems (Moran, 1997).

If mathematics teachers allow group work, discussion, and information gathering in libraries and over the Internet, and also want students to learn more mathematics in collaborative work, then they face great demands on what types of problems they should pose. Silver and Kilpatrick (1989) argue for the use of open-ended problems in the assessment of mathematical problem solving, thereby moving from facts and procedures to concepts and structures. A relevant problem should encourage students to make various assumptions and use various strategies in which technology can serve as an aid but never as a goal. The problem teachers choose also need to provide the students with opportunities to express what they have learned in the course and in previous courses. At the same time that the problem should remain nontrivial in the presence of technological tools, their use should not be the only performance component that is essential and leads to success (Lingefjärd and Holmquist, 2001).

Bliss (1994) distinguishes between learning about a mathematical model that someone else has constructed through exploration and the process of building a mathematical model through which learners can express their own understanding of a situation. Two distinct but inseparable kinds of learning are implied. In the first case, learning about someone else's mathematical model involves appropriation of the meanings embedded in the mathematical model's context by the creator of the model. In the second case, creating a computer-based model on one's own suggests that the learner will generate meaning as she or he actively engages in externalizing her or his understanding through the process of mathematical model building using the technical resources at hand.

3. Students' work

We now discuss the way a group of prospective teachers at Göteborg University dealt with a mathematical modeling situation, how the students developed mathematical models, and how their work was assessed. The 11 students in a mathematical modeling class in spring 2001 were in their third year of mathematical studies, preparing to become teachers of mathematics and of a complimentary subject such as physics, chemistry, or biology in the Gymnasium (Grades 10 to 12). All had taken courses in number theory, Euclidean geometry, linear algebra, real analysis, probability, statistics, and discrete mathematics. Especially through their work in the discrete mathematics course, all of the students had some collaborative experience in solving extended problems requiring approaches unlike those usually practiced in class or in using other mathematical literature than the normally required textbook to support their arguments. The mathematical content of the course was designed to give the students insight into how they could solve extended problems using mathematical modeling by drawing on technology and their background in mathematics. The software used was mainly The Geometer's Sketchpad (Jackiw, 1995), Derive (Texas Instruments, 2000), Excel (Microsoft, 1997), and CurveExpert (Hyams, 1996). Graphing calculators were available. The students worked in a computer lab with all the software described above and with access to the Internet.

The final assessment was arranged over a working period of three weeks, and each student had to work out, solve, and hand-in a written report on three different extended modeling situations. Based on the assumptions and perspectives discussed above, the following situation was presented as part of the final assessment during the spring 2001. It was formulated and presented as an investigational problem as it includes the students' own creation of some data.

A modeling situation (from Edwards and Hamson, 1996, pp. 110-111, with our additions and modifications)

Figure 1 represents a small lake. Although the lake is receiving and losing water in different ways, we simplify the situation and say that water flows in through stream A and out through stream B.

At a certain time of the day, as a result of a road accident, a petrol truck overturns and spills a toxic chemical into the stream A at position X.

Thirty minutes later the police and emergency services have brought the situation under control, and an unknown amount z (m^3) of the toxic chemical has leaked into the lake.

Develop a mathematical model that you can use to predict the concentration of the pollutant in the lake at any time and use it to estimate (for a range of possible initial pollution amounts z):

a)   The maximum pollution level in the lake and the time at which the maximum is reached.

b)   The time it will take for the pollution to fall below the safe level of 0,05%.

c)   How will your results be affected if a constant rain starts at the same time as the accident? The rain covers the whole geographic area.

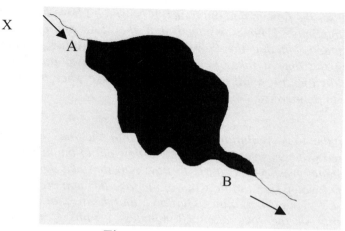

Figure 1

4. Students' responses

It should be stressed that naturally we were unable to investigate the learning process in a direct way; we could only try to observe the learning outcomes in terms of external characteristics such as students' behaviors, attitudes, and skills. It should also be mentioned that the instructors' view of learning and assessment converged in the way the solutions of the problem were examined. The two instructors examined each of the students' papers, and most of the papers were then discussed with the students. This discussion took place with the students in groups or in pairs and underlined the fact that the examination was integrated with the process of learning. Some of the students had to rewrite their papers to pass the exam, some had to take a new exam, and some gave a short summary of how they had solved the problem.

Within the class of 11 students, a smaller group of 6 students had formed to work together. The discourse of the smaller group took on a tone of authority, which led the students in the group to follow one or possibly two "strong" students when developing their mathematical model for the exam problem. They thereby abandoned the possibility to develop their own model and instead chose to investigate another student's model. The following three quotations from students' papers illustrate the results from that group.

S 1

I have set the flow in and out to 0,1 m^3/s, which I have transformed into 6 m^3/min for simplicity. I have assumed that the flow out of the lake will remain unchanged, but that the flow into the lake increases somewhat (z/30 m^3/min) during the time it takes for the toxic pollution to pour into the lake. I have calculated with a small enlargement of the lake when the pollution flows in, but not with an increased outflow.

S 2

When the toxic pollution has poured into the lake through the creek A, no polluted water has yet had the time to flow out of the lake through creek B. After 30 minutes fresh water starts to flow into the lake again through creek A and polluted water starts to flow out of the lake through creek B. My idea is that the Inflow = 6 + z/30 m^3/min = Outflow...that means that it is 19 m^3 toxic chemical pollution in the lake after 30 minutes. I think it is reasonable, since my assumption was that 20 m^3 had leaked out of the petrol truck. The missing m^3 have probably disappeared between X and A, or in the lake.

S 3

Inflow = 6 + z/30 = Outflow. I get the amount by adding the original flow of water to and from the lake with the toxic leak from the petrol truck spread out over the thirty minutes. Since I have chosen to keep the volume of the lake constant, I get that the Outflow also increases with z/30.

The first two student responses express a view that is based on a doubtful interpretation of how flow and water level are affected by the discharge. Several of the students in the group described the alteration as a "tidal wave," an additional amount of fluid that moves like a pulse through the system. That in turn results in difficulties when the mathematical model and its behavior are evaluated. Since this line of reasoning was common in the group, we see it as a result of the discussions that had taken place among students in the group during the modeling process. The discourses the students have been part of were dominant in a negative way for the individual student. The third student (S3) is one of the few who expressed a more consistent line of reasoning according to assumptions made in the modeling process. That approach may have been connected to the instructors' observation that this student was one of the strong students in the group.

In any student response to open mathematical modeling problems, it is expected that several components of that student's view of mathematics become visible. Fischbein (1994) discusses three different components or aspects of mathematics as a human activity: the formal aspect, the algorithmic component, and intuition. His statement is focused on the comprehension of mathematical theories, but he also discusses the perspective of problem solving and one's ability to draw reasonable conclusions from a mathematical line of reasoning. In the mathematical modeling process the formal aspect could be seen as the student's knowledge of required mathematical theories, the algorithmic component as the skills needed in the solving procedures, and the intuition component as the possibility for the student to validate a complicated mathematical model. There are obviously both interactions and conflicts between these aspects and components in a student's mathematical modeling activity. There is also a risk that some students could work too hard in their ambition to make their mathematical model "complete," thereby meaning that it should describe and control every single detail of the pollution accident:

S 4

The factor that affects the pollution in the lake the most is the size of the flow of toxic chemicals from the petrol truck, which in turn is related to the cross-sectional area of the aperture.

The student who wrote this statement is considered to be a fairly good mathematical student, who in a mathematically correct presentation constructed her model on a fairly sophisticated level. Her detailed model of how the geometrical properties of the aperture in the petrol truck direct the pollution of the lake prevented her from making a free validation of the model. She also divided the lake into 18 discrete zones with a certain cross-sectional area. Her theoretical framework finally resulted in the following statement about the concentration of pollution in the lake:

S 4

The concentration is proportional to the cross-sectional area, which yields the highest concentration at the inflow and outflow positions in the lake.

The aim for a detailed mathematical model that has a one-to-one correlation with reality, and the search for a "general" algorithmic solution apparently annihilated this student's intuitive interpretation. This phenomenon could be compared to the following summary by another student in the course, who expressed a more balanced view when looking back on how the results of the mathematical model would be affected if a constant rain starts:

S 5

If the inflow is larger than the outflow, it will result in a rise in the surface level in the lake. What this actually means is very hard to say. Will it cause the surface of the water to attract more chemical and if so – how much? What is the lake's potential to store more water? When the surface level rises, it might very well result in new streams of outflow. When I validate my mathematical model against all these assumptions, I realize that my mathematical results may be exceedingly doubtful.

5. Conclusions

We believe that the reflections that students make about on their activities and about their personal processes of mathematization, as defined by de Lange (1996), are strongly affected by the discourse that is developed in a certain student group. One important aspect of an individual student's behavior in such a discourse is the student's self-confidence in his or her knowledge of mathematics and its applications in the modeling process. As we have discussed elsewhere (Lingefjärd and Holmquist, 2001), it also seems to be easy for the students to "get lost" and trust the technology far too much, thereby avoiding a necessary validity check.

Another important conclusion from the course work is that even when students solve a mathematical modeling problem using data points generated from their "own model," they can indeed become confused. Sometimes misled by their desire to come up with sophisticated solutions, they become blind, unable to see simple, elegant solutions.

The formal and intuitive component of a student's view or understanding of mathematics can also be applied to the modeling process. According to Fischbein (1994), there are situations in which the intuitive understanding stops or otherwise disturbs formal understanding. Nevertheless, we consider that we have seen the opposite – the formal component of the students' mathematical ability sometimes creates an obstacle for the intuitive interpretation of the situation, the process that is modeled. It is obviously essential to find a good balance between solid formal knowledge and insight in how to use this formal knowledge and how to develop a good intuitive capacity. This is especially important for prospective teachers, who should be trained to guide their own students towards the goals, in particular concerning mathematical modeling, that are proposed in the curriculum and syllabus in mathematics.

However, there are several other benefits we teachers can hope for when presenting students with investigational problems like the one described in this article – the enthusiasm and eagerness to find solutions that some students will show, the fact that different students will come up with different solutions and share these with each other, and the fact that they will develop a deposition to notice and enjoy all the mathematics that is embedded in the world around us.

References

Bliss J (1994) 'From mental models to modelling' in Mellar H et al (Eds) *Learning with artificial worlds: Computer-based modelling in the curriculum,* London: Falmer Press, 27-32.

Clarke D (1996) 'Assessment' in Bishop AJ et al (Eds) *International handbook of mathematics education,* Dordrecht: Kluwer, 327-370.

De Lange J (1996) 'Using and applying mathematics in education' in Bishop AJ et al (Eds) *International handbook of mathematics education,* Dordrecht: Kluwer, 49-97.

Edwards D and Hamson M (1996) 'Mathematical modelling skills', London: Macmillan.

Fischbein E (1994) 'The interaction between the formal, the algorithmic, and the intuitive components in a mathematical activity' in Bielhler R et al (Eds) *Didactics of mathematics as a scientific discipline,* Dordrecht: Kluwer, 231-245.

Hyams D (1996) 'CurveExpert: A curve fitting system for Windows' Clemson, SC: Clemson University.

Jackiw N (1995) 'The Geometer's Sketchpad', Berkeley, CA: Key Curriculum Press.

Lingefjärd T (2000) 'Mathematical modeling by prospective teachers' Electronically published doctoral dissertation, University of Georgia. Can be downloaded from http://ma-serv.did.gu.se/matematik/thomas.htm

Lingefjärd T and Holmquist M (2001) 'Mathematical modeling and technology in teacher education - Visions and reality' in Matos J et al (Eds) *Modelling and Mathematics Education ICTMA 9: Applications in Science and Technology,* Horwood: Chichester, 205-215.

Marton F and Booth S (1997) 'Learning and awareness', Mahwah, NJ: Erlbaum.

Microsoft (1997) 'Microsoft Excel', Stockholm: Microsoft Corporation.

Moran JJ (1997) 'Assessing adult learning: A guide for practitioners' Malabar, FL: Krieger Publishing.

Niss M (1993) 'Assessment of mathematical applications and modelling in mathematics teaching' in de Lange J et al (Eds) *Innovation in mathematics education by modelling and applications*, Chichester: Ellis Horwood, 41-51.

Ramsden P (1992) 'Learning to teach in higher education', London: Routledge.

Schoenfeld AH (1992) 'Learning to think mathematically: Problem solving, metacognition, and sense making in mathematics' in Grouws DA (Ed) *Handbook of research on mathematics teaching and learning*, New York: Macmillan, 334-370.

Sfard A (1997) 'From acquisitionist to participationist framework: putting discourse at the heart of research on learning mathematics' in Lingefjärd T and Dahland G (Eds) *Research in mathematics education* (A report from a follow-up conference after PME 1997, Report 1998:02, pp. 109-136), Gothenburg: Gothenburg University, Department of Subject Matter Didactics.

Silver EA and Kilpatrick J (1989) 'Testing mathematical problem solving' in Charles R and Silver E (Eds) *Teaching and assessing mathematical problem solving*, Hillsdale, NJ: Lawrence Erlbaum, 178-186.

Skolverket (2000). *Kursplaner och betygskriterier för kurser i ämnet matematik i gymnasieskolan.* [The Swedish secondary school curriculum and syllabus for mathematics]. Electronically published document. Stockholm: Skolverket. Can be downloaded from http://www.skolverket.se/styr/index.shtml

Sternberg RJ (1996) 'What is mathematical thinking' in Sternberg RJ and Ben-Zeev T (Eds) *The nature of mathematical thinking*, Mahwah, NJ: Lawrence Erlbaum Associates, 303-318.

Texas Instruments (2000) 'Derive 5', Stockholm: Texas Instruments InCorporation.

18

Two Modelling Topics in Teacher Education and Training

Adolf J I Riede
University of Heidelberg, Germany
riede@mathi.uni-heidelberg.de

Abstract

The focus of this chapter lies particularly in the practice of the didactical capacity of two aspects of modelling methodology: Studying critically several models of the same problem and modifying a model to deal with a different situation. One topic deals with models to compare performances in sport. The other is concerned with models of systems approaching a saturation point.

1. Introduction

The report is based on two seminars at the University of Heidelberg for students intending to become high school teachers and on a high school teacher-training course for one week in the Mathematical Research Institute in Oberwolfach, Germany. Part of the program was carried out in conjunction with a high school teacher. One case study deals with new insight into models to compare the performances of weightlifters of different weight classes, high jumping with respect to different body length, the decathlon scoring system and running short, middle and long distances. In the other example the logistic and another model of systems approaching a saturation point are analysed with regard to teaching in the classroom.

2. General Didactical Aspects

This paper promotes two ideas of modelling methodology and didactical philosophy. Firstly, teaching how to apply mathematics in private, professional and social life is important for good citizenship and in professional life even in non-mathematical careers. Secondly mathematical modelling can be used to promote mathematical understanding beyond the procedural use of formulaic approaches to solving problems.

In this article the focus lays particularly in the practice of the didactical capacity of experiencing several models. The critical study of different models of the same problem can be used to promote mathematical comprehension. As the focus of these seminars was on mathematical modelling appropriate to secondary school programs the examples recorded in this paper will reflect this emphasis.

3. Comparison of Performances in Sports

3.1 Weightlifting

3.1.1 The problem There are two principal lifts namely the snatch and the clean and jerk. We consider here only the snatch where the weight is pulled from the floor to a locked arm overhead position in a single move although the lifter is allowed to move or squat under the weight as it is being lifted.

The problem is how to compare the performance of weightlifters of different weight classes. This problem was discussed by Burghes et al (1982) and used by Houston (1993) in a comprehension test.

Let us first consider the world records L in the different weight classes B up to December 31st 2000 inclusive. This date is taken because the associations of weightlifting use a model that is based on the world records until the end of an Olympic year and this is used until the end of the next Olympic year, i.e. for one Olympiad. The world records are posted in the Internet (www.sport-komplett.de or www.iwf.net). The bodyweight of an athlete in the weight class up to B can be put equal to B because in practice it is very close to the upper boundary of the weight class. In the highest class the actual body weight B of the athlete is taken on the day when she or he achieved the record. This can be found in the database of the Institut für Angewandte Trainingswissenschaft, Leipzig (www.iat.uni-leipzig.de/weight.htm).

In 1998 a new definition of weight classes was introduced and the world records in the new classes up to 1998 were represented by the so called world standards. In the men's class 94 < *B* 105 kg this world standard was not achieved by an athlete before the end of 2000.

Ladies Snatch World Weightlifting Records up to 31-December-2000

B [kg]	*L* [kg]	Name	Country	Date	Town
48	87.5	LIU Xiuhua	CHN	09-Jun-00	Montreal
53	100.0	XIA Yang	CHN	18-Sep-00	Sydney
58	105.0	CHEN Yanqing	CHN	22-Nov-99	Athens
63	112.5	CHEN Xiaomin	CHN	19-Sep-00	Sydney
69	112.5	MARKUS Erzsebet	HUN	19-Sep-00	Sydney
75	116.0	TANG Weifang	CHN	03-Sep-99	Wuhan
103.56	135.0	DING Meiyuan	CHN	15-Sep-00	Sydney

Table 1

Mens Snatch World Weightlifting Records up to 31-December-2000

B [kg]	*L* [kg]	Name	Country	Date	Town
56	138.0	MUTLU Halil	TUR	16-Sep-00	Sydney
62	152.5	SHI Zhiyong	CHN	03-May-00	Osaka
69	165.0	MARKOV Georgi	BUL	20-Sep-00	Sydney
77	170.5	KYAPANAKTS-YAN K.	ARM	25-Nov-99	Athens
85	181.0	ASANIDZE Georgi L.	GEO	29-Apr-00	Sofia
94	188.0	KAKHIASVILI S Akakios	GRE	27-Nov-99	Athens
105	197.5	World Standard		01-Jan-98	
147.48	212.5	REZAZADEH Hossein	IRI	26-Sep-00	Sydney

Table 2

3.1.2 The comparison procedure To compare the lifts of athletes of different weight classes we interpret the procedure of Burghes et al (1982) as follows: There is a function $F = F(B)$ of body weight B, which describes the performance capacity F of an athlete of bodyweight B. This function is found approximately by fitting a function of a suitable type to the world record data.

To model the dependence of the performance on body weight, we try three functions.

(a) The regression parabola for the female and the male case are:

$$F = 0.00923\ B^2 + 2.173\ B + 8.247,$$
$$F = 0.00915\ B^2 + 263\ B + 22.74$$

(b) A model by a power function calculated by logarithmic regression:

$$F = 13.0\ B^{0.51}\text{ in the female case, and}$$
$$F = 25.5\ B^{0.43}\text{ in the male case}$$

(c) Mainly for older athletes the sport associations use the Sinclair total ST and the Sinclair coefficient SC. In the sense of this paper they are based on a model which is called the Sinclair Model and is used for one Olympiad.

$$F = L_m/G_m \text{ with } G_m = 10^{X_m} \text{ and } X_m = A\ (\log_{10}(B/B_m))^2$$

Here L_m is the world record of the heaviest athletes and B their body weight on the day when they achieved the world record. The function is adjusted in such a way that the value for the heaviest world record athlete is $L = L_m$. (Incidentally $SC = L_m /F(B)$ and $ST = L$.) We have determined the coefficient A by regression from the logarithmic data that turned out to be a linear regression problem:

$$A = 1.7094 \text{ in the male case and } A = 1.0684 \text{ in the female case.}$$

The graphs in the Figure 1 and Figure 2 suggest, at least in the male case, the parabolic model offers the best fit, followed by the Sinclair model and then the power function.

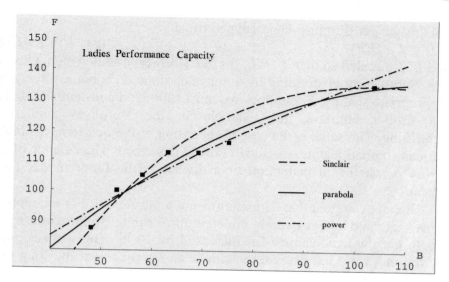

Figure 1

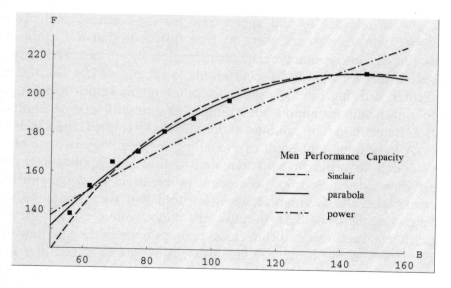

Figure 2

From Burghes et al's (1982) definition of the handicapped lift we deduce that there are two possibilities to define the handicap H:

(a) $H = H(B) = F(B) - F(75)$ or   (b) $H = H(B) = F(B) / F(75)$

From this we get Burghes' handicapped lift L':

(a) $L' = L - H(B)$ or   (b) $L' = L / H(B)$

H and L' are scaled so that $L' = L$ for the (former) middleweight class of 75 kg. Subtracting or dividing by a constant alone does not always yield an appropriate scaling. Sometimes we must subtract a constant and divide by a further constant, for example in the Vorobyev model of weightlifting. The same is the case in the rating of the quarterbacks in the American football league, as Burrill (2001) described. The essence of the question is whether a deduction or a divisor is the fairer model for a handicap?

It can be argued that deduction is not a fair model: For example, if a lightweight and a heavyweight athlete improve their performance by 10 kg, both get the handicapped lift increased by 10 kg. But it takes much more training for a lightweight to achieve an improvement of 10 kg than for a heavyweight.

3.1.3 Teachers' remarks   The participating teachers believed that these models were relevant to and appropriate for high school students but questioned why it was necessary to find functions that did more than simply linearly interpolate the data or interpolate by splines. The simplest response is that we want to have a formula to calculate $F(B)$. Considering this further, a fitting function smoothes out singular values which occur due to significant variation - for example an extraordinary performance like Bob Beamon in long jumping at the Mexico Olympic Games in 1984.

A very important point of modelling methodology is certainly to apply models to different situations and to find new applications. One teacher directed the attention to scoring problems in team weightlifting for example in the German leagues. He told that they might be very interested in these models because the clubs often do not have a weightlifter in each weightlifting class; perhaps none in the middleweight class but two in the heavy weight class. A report on modelling team weightlifting can be found in section 3.1.4. Another teacher mentioned the decathlon where a points system is also being used. The decathlon scoring system will be explained in section 3.2.

3.1.4 The relative deduction point table and an alternative   Formerly the total weight of a weightlifting team was restricted to 550 kg (the "11 quintal rule") and the lifts of all members of a team were added to get the team's result. Of course this was not very satisfactory and often a team had to loose up to 20 kg within a few days before a competition.

Since 1990, every athlete is given, according to his body weight, a deduction of his lift L, the so-called relative point deduction D. The relative points of the athletes are calculated by the formula $P = L - D$ and relative points of all members of each team are summed to give the team's score. In practice the German weightlifting association use the relative deduction point table. Here is a short extract for the male case:

B	52	56	60	67.5	75	82.5	90	110
D	36	42	50	63.5	72.5	82.5	90	104

Table 3

The relative deduction table is a good fit to the performance capacity. A calculation shows that for the men's world record in the snatch event up to 31 July 1977 we get nearly the same ranking as by the Russian Vorobyev model as well as by O'Carrol's on which Houston reported (1993).

On the other hand it is inaccurate, for example, in the men's table between 79.1 kg and 95.5 kg body weight because the deduction equals the bodyweight. This would be correct if the performance capacity grew linearly with increment 1 in this body weight range. But this is not the case.

During a recent competition I asked a number of fans of our local weightlifting club to explain the relative deduction points system. They knew that they are based on the world records but did not understand the meaning of the points. They were happy with the scoring system, however, perhaps because of the completely unsatisfactory situation that existed before the relative deduction table was created.

Furthermore, I have argued above that there is some doubt that a process based on deduction is a fair method of handicap. Alternatively we can use a process based on division. Let us call $F(B)$ *the average world record for the bodyweight B*. Take the ratio of the lift L to $F(B)$, multiply by 100 and take the result as say ratio points (instead of deduction points). Then 80 ratio points mean that the athlete has achieved 80% of the world record for his bodyweight. The world record for a bodyweight is an average based on actual world records. This system of points would overcome the above-mentioned problems and the ratio points would have a meaning that every weightlifting fan could well understand. The averaging method could be explained to the layman by showing a graphic like the above (of course with only one graph). There still remains the question, though, of which model to use.

3.1.5 Power law, Vorobyev, parabola model It is known that the performance of the heavy weight class does not fit well with a power law model. A reason could be that the heavy weight class has relatively few competitors. But O'Carroll's power law model was statistically population-compensated and we see nevertheless that the heavy weight - and especially the super heavy weight- does not fit with a power law model. The super heavyweight can only be considered if we know the athlete's body weight at the day of the record.

The Vorobyev model, on the other hand, takes into account that the competitor also lifts his own body and therefore adds the bodyweight of the athlete to his lift and further ideas are realized. Anyway the result is that Vorobyev's degree of merit suits the idea of a performance capacity function. Furthermore the Vorobyev's performance capacity function is a parabola function!

Why can a parabola be a good model? An athlete's absolute strength generally increases with weight. But increased weight also makes it more difficult to perform the movements that are necessary in weightlifting. The complexity of this movement is comparable with that of pole-vaulters. So performance must decrease from a certain body weight onwards. The Vorobyev model is structured such that an athlete of body weight 465 kg has zero performance. Obviously it is assumed that from a bodyweight of 465 kg upward the complicated movement can not more be accomplished. The good fitting properties of a parabolic function favours the parabola model, which on the other hand is a very easy model. For example for a polynomial function linear regression can be done whereas the coefficients in the Sinclair model cannot be determined by linear regression.

Two questions that arise from the above discussion are:
(a) Can the assertions on the parabola model be confirmed by calculations?
(b) Can we predict in which female bodyweight classes the world record is most likely to be improved soon? Would this be consistent with the new world record that was recently achieved in the class up to 69 kg: 113.5 kg, POPOVA, Valentina, RUS, 1.9.2001, Brisbane?

3.2 Decathlon and running short middle and long distances

De Haan and Wijers (2003) reported at ICTMA 10 about a modelling competition organized by the Freudenthal Institute in Utrecht. One of the assignments within this competition was to describe how points were awarded in the decathlon and the women's heptathlon. It is interesting to compare the scoring systems. Instead of using percentages they count in the combined events by thousands; 1000 points are given for the highest performance, 0 points are given for an assumed lowest performance, for example in the pole vault today for a 1 m jump. On the other hand a 1 m jump corresponds to 16 ratio points because 1 m is approximately 16% of the world record 6.14 m. Similarly 0 m corresponds to 0 ratio points. In the early nineties the dependency of the points on the performance between the lowest and highest scores was modelled as a linear function of the speed in the track events and as a linear function of the square root of the distance in the field events. This has changed now to the following formulae

$$\text{Points} = a(b - M)^c \quad \text{in the track events}$$
$$\text{Points} = a(M - b)^c \quad \text{in the field events}$$

Here M is the time in sec in the track events, distance in cm in the jumping events and distance in m in the throwing events. b is the performance that corresponds to 0 points, a and c are further constants depending on the event and are different for men and women. This change leads to the question: Should we take time or speed as measure for the performance in the running events? There seems to be some advantage taking the speed rather than the time. For example Schober (1995) got a high school class to model the dependency of the performance in track events on the distance of the event. If you measure the performance as time you must use a strongly increasing model function. But if you take the speed as measure for the performance then a monotone function is not appropriate for the short distances because the average speed of a 200 m sprinter is higher than that of a 100 m sprinter due to the effect of the starting procedure. This may correspond to Schober's students' observation that their power law model did not fit well for short distances.

3.3 High Jumping

Shorter students often avoid taking part in high jumping because they know in advance that their taller classmates will in most cases jump higher. With this in mind it seemed a worthwhile exploration to model the performance capacity for high jumping in relation to the body length of an individual. This investigation was based on data collected from my athletics sports group. This is known as the "everybody-sports-group" because everybody can participate. The data is not representative of the general population but authentic. This data was used with a student who was asked to formulate a linear model and to use this to answer the question "If a person is 10 cm taller how much higher can the person jump in average?" The first task was to search out a comparable, suitable subgroup. I then asked him to write a report for the sports-group.

The data for the subgroup is presented below. The unit is one meter. l = body length, h = the height jumped.

l	1.84	1.75	1.90	1.89	1.80	1.80	1.74	1.74	1.75	1.80	1.75	1.80
h	1.05	1.10	1.25	1.10	1.25	1.15	1.15	1.25	1.15	1.15	1.15	1.10
l	1.85	1.84	1.79	1.72	1.75	1.66	1.66	1.72	1.75	1.77	1.70	1.58
h	1.15	1.05	1.05	0.90	1.10	1.05	1.10	1.20	1.10	1.10	1.05	0.90

Table 4

And here are two straight lines fitted more or less to the data.

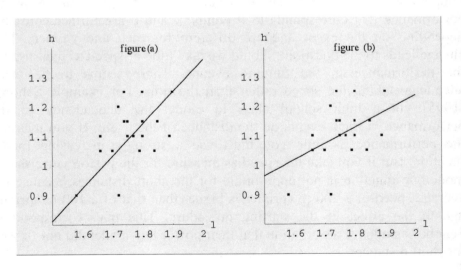

Figure 3a   Figure 3b

$h = 1.094\, l - 0.828$ in figure (a), $h = 0.546\, l + 0.146$ in figure (b).

One line is calculated by regression. The reader should first try to guess which of the straight lines has been found through calculation.

Here is an extract from the student's written report for the everybody group:

"...The value 35/32 is constructed in the following way. 32 cm is the difference between the shortest body length and the longest. 35 cm is the difference between the lowest height jumped and the highest. From this I calculated the increment of 35/32 (= 1.094). It can be seen from the standard deviation q whether this straight line is a good approximation to the regression straight line. For my straight line (a) this is $q = 0.977$ and for the regression straight-line (b) it is $q = 0.078$. This means that my straight line is not a good approximation because these values differ a lot. From the figures (a) and (b) I could not state this. Rather I had the impression that the straight line (a) fits better than (b). ..."

I observed that it is not easy for students to write a report. In this case he had to learn a lot about regression and do calculations using a Computer Algebra System. In consequence I required him to answer 8 questions as part of his report. These questions appeared to assist the student as they provided focus for the report.

The regression straight line in Figure 3(b) has increment 0.546 and standard deviation of approximately 0.08. For the everybody group this means: A person who is 10 cm taller can jump on average 5.5 cm plus or minus 0.8 cm higher. Although the data seemed to be very dispersed, this is a fairly sharp answer appropriate to the target group, which consisted of elder (male) hobby sportsmen. A question worthy of further investigation is how generally applicable this process could be used - for example to a group of 15 years old schoolboys or schoolgirls? Further information is available from a platform on the Internet where by using the keyword "Hochsprung relativ" further pedagogical ideas can be found that are relevant to organising high jumping in school sports.
(www.sportpaedagogik-online.de/leicht/larelativ.html)

4. Approaching a Saturation Point

Approaching a saturation point S -for example the maximum length of a plant- has a nice modelling procedure. Let x denote the size of the plant. We look at the dynamical system for exponential growth:
$$x_{n+1} = qx_n.$$
Replace the constant growth factor q by a function $q(x)$ that is decreasing and satisfies $q(S) = 1$. The easiest decreasing functions are $q(x) = -bx$ and

$q(x) = a/x$ with b and a positive constants. Then translate x by a suitable constant to obtain $q(S) = 1$. We get:

Logistic law

$$q(x) = -b(x - S - 1/b), \quad x_{n+1} = (bS + 1)x_n - bx^2 = ax_n - bx_n^2 \text{ with } a = bS + 1$$

The rational law (Let us give it this name for the purpose of this article.)
$$q(x) = a/(x + a - S) = a/(x + b) \text{ with } b = a - S, \quad x_{n+1} = (ax_n)/(x_n + b)$$

Assume $a > S$ such that the model renders positive values.

These are two possible types of functions and others can be constructed in a similar manner. The following exercise is offered in order to further stimulate interest in the modelling process: In a dynamical process where according to the data the variable approaches a saturation point determine which law provides the better fit for the data. To keep the calculations within bounds let us test for weekly rather than daily development. Our first week starts with the 7th day because the data for the early days have a larger relative error.

Day	1	2	3	4	5	6	7	8	9	10
Height	3.3	3.9	4.7	5.7	6.8	8.1	9.7	11.6	13.8	16.5

Day	11	12	13	14	15	16	17	18	19	20
Height	19.5	23.1	27.2	32	37.4	43.6	50.4	58	66.3	75.2

Day	21	22	23	24	25	26	27	28	29	30
Height	84.7	94.6	104.7	114.9	125	134.9	144.4	153.4	161.7	169.3

Day	31	32	33	34	35	36	37	38	39	40
Height	176	182	188	193	197	200	203	206.1	208.3	210

Day	41	42								
Height	211.8	213.1								

Week	1	2	3	4	5	6
Height	9.7	32	84.7	153.4	196.7	213.1

Table 5

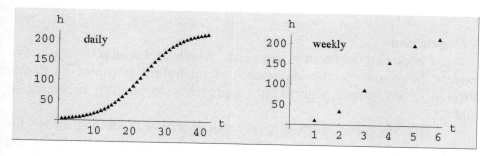

Figure 4

A logistic and a rational model are easily found by fitting exactly to the first three data points. This only requires a pair of linear equations to be solved in each case. Thus this should also be applicable to high school programs. We get $a = 3.58$, $b = 0.029$ in the logistic model and $a = 298.6$, $b = 80.8$ in the rational model. The following figures show which of the models fits better. The figure on the left shows how the logistic model fits the original data. By construction it fits exactly the first three points. The figure on the right shows that the rational model describes very precisely the original data. This example may help teachers to involve their students more in the modelling procedure and thus promote a deeper understanding of Mathematics. It motivates consideration of a more open problem rather than merely finding parameters for a logistic model.

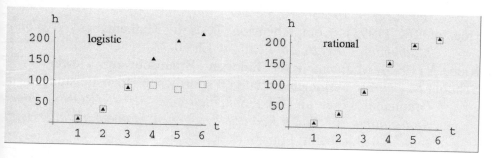

Figure 5

The clue to this exercise does not appear to be well known. The data are discrete values of a solution of a logistic differential equation. Then it is a theorem that these discrete data satisfy a rational difference equation and not a logistic one as can be seen in Riede (1993), p 150f.

5. Conclusion

I hope that the ideas in this paper, which deal with several types of models, help promote the importance of modelling as a process that is different from simple algorithmic approaches to solving problems in context.

The context of sports is a very rich source of modelling ideas - for example, in an event like power lifting where the so-called Siff model and others are used. The Siff model approaches a saturation point but that is not the case in weightlifting as we argued! For schools and for universities there are scoring tables for the assessment of performances in sports that provide a great range of opportunity for study and investigation.

Acknowledgements

The author thanks J. Roy Dennett, Hull and Vincent Geiger, Adelaide for advice in the preparation of this chapter.

References

Burghes DN, Huntley I and McDonald J (1982) 'Applying Mathematics', London: Ellis Horwood.

Burrill G (2001) 'The NCTM Standards 2000 - Guidelines for Mathematics Teaching in the Information Age' in Kaiser G (Ed) *Beiträge zum Mathematikunterricht* 2001, Hildesheim: franzbecker, 19-26.

Houston SK (1993) 'Comprehension Tests in Mathematics' *Teaching Mathematics and its Applications* 12, 60-73.

Riede A (1993) 'Mathematik für Biologen', Braunschweig: Vieweg.

Schober M (1995) 'Vergleichbarkeit von sportlichen Höchstleistungen' in Graumann G et al (Eds) *Schriftenreihe der ISTRON-Gruppe: Materialien Für Einen Realitätsbezogenen Mathematikunterricht, Band 2*, Hildesheim: Franzbecker, 15-21.

Section F

Innovative Modelling Courses

19

The Knowledge and Implementation for the Course of Mathematical Experiment

Zhao Jing, Jiang Jihong, Dan Qi and Fu Shilu
Basic Department, Logistics Engineering College, Chongqing, PRC.
zhaojing.@163.com

Abstract

In this paper, the research and practice for the course of mathematical experiment in Logistics Engineering College is shown. The main content is as following:
1. The Knowledge for the Course of Mathematical Experiment
2. The Content and Implementation of Mathematical Experiment
3. Examples
4. The Effects of the Course of Mathematical Experiment

The 21st century needs creative persons of high quality. Mathematical science can inspire one's intelligence and stimulate one's creative ability. So mathematical quality is an important character for creative personnel. Mathematical Modelling and Mathematical Experiment are important methods to realize highly mathematical quality education.

1.The Knowledge for the Course of Mathematical Experiment

Most people know the importance of mathematical education in the whole personnel training process. A traditional mathematical education system focuses on definitions, themes, formulae and calculations. It lays stress on training students to study known knowledge. We all know that strict logical inference, accurate and fast calculation are important for the students who will solve all kinds of real world problems with the tools of mathematics. But, for a real world problem, one must describe it with mathematical language and form its mathematical models at first, then solve the mathematical models. At the end he must analyze, examine and refine the solutions by practice. Unfortunately, the training for such ability of "Applying Mathematics" is deficient in the traditional mathematical education. In recent years, the activity of Mathematical Modelling, being developed in the universities, is a good attempt and has achieved good results. On the basis of Mathematical Modelling activity, giving the course of Mathematical Experiment is a new attempt to improve the mathematical education system, teaching content and teaching method for enhancing the ability of "Applying Mathematics" of students.

Mathematical experiment is a new object arising by introducing computers and mathematical software into mathematical education. In this course, the main task of students is "studying mathematics" through "doing mathematics". With the help of computers and mathematical software, students, by themselves, calculate, sample, examine, research and solve. This could strengthen their understanding and interest to mathematics, develop their ability to solve real world problems using mathematics, arouse their enthusiasm to study mathematics and promote the virtuous circle of mathematical education.

The idea of offering the course of Mathematical Experiment appeared in China from the middle of 1990's and it evoked great repercussions in the university mathematical education area. In 1997, the National Instruction Committee for Engineering Mathematical Courses in Colleges and Universities suggested that Mathematical Experiment should be one of the foundation mathematical courses. Following this suggestion, Logistics Engineering College established the Mathematical Experiment Laboratory in 1997 and presented the course of Mathematical Experiment from 1998. Through two years' researching and practice, we made Mathematical Experiment a required course in our university from September of 2000.

2. The Content and Implementation of Mathematical Experiment

Mathematical Experiment is a new kind of course and has many problems to consider. In our opinion, its main teaching aim is to strengthen students' interests in mathematics and develop their ability of "Applying Mathematics". Through doing mathematical experiments, students understand deeply basic concepts and basic principles of mathematics, master proficiently the common mathematical software, develop the ability to build mathematical models and solve real world problems using a computer. The emphasis of our Mathematical Experiment course is to develop students' research spirit for science. So in this course, we must endeavour to train students to examine, to research and to solve. Guo Xibo et al (1999) wrote that "investigating" means getting students to implement the mathematical theories and methods on a computer by themselves to strength understanding. "Researching" means that through completing highly accurate and enormous calculations using a computer, students will find some truths, which are difficult to find by logical inference. "Solving" means students solve mathematical models quickly with a computer. We must arrange the teaching content closely around the basic aim and emphasis of Mathematical Experiment course.

In our university, we combine mathematical experiment with the course of "Engineering Mathematics" (including 210 class hours of Higher Mathematics and Linear Algebra) and "Mathematical Modelling". According to the varying requirements of these two courses, we divide mathematical experiment into two parts, "Engineering Mathematical Experiment" and "Mathematical Modelling Experiment".

"Engineering Mathematical Experiment" is given in step with the theory course of "Engineering Mathematics" in the first year for undergraduates. The goal of this course is to let students understand the theory of "Engineering Mathematics" more deeply with the aid of mathematical software and to develop students' interests in mathematics and the consciousness and ability to use mathematical software to solve mathematical problems. The focal points are "investigating" and "researching". According to the teaching content of "Engineering Mathematics", we arrange 12 experiments and give 20 class hours for the course of "Engineering Mathematical Experiment". Six experiments are carried out separately in the first semester and the second semester. In each experiment, we firstly teach students how to solve mathematical problems with software. Then we guide students to observe and learn mathematical appearance from experiment, or teach students how to build mathematical models for simple real world problems and solve the models by software. At last, we give some experiment tasks for students to do after class.

"Mathematical Modelling Experiment" is given in the course of "Mathematical Modelling" in the second year for undergraduates. The goal of this course is to develop students' ability to build mathematical models and solve models with mathematical software, that is, the ability of "Applying Mathematics". The focal points are "researching" and "solving". The contents of an experiment consist of five topics: optimization method (linear programming, nonlinear programming), differential equations, graph theory, mathematical statistics and data processing (parameter estimation, hypothesis testing, regression analysis, interpolation, fitting) and computer simulation. We have 30 lecture hours for the course of "Mathematical Modelling Experiment". In each experiment, we simply introduce the applied mathematical knowledge most in use to solve real world problems, and how to use the corresponding applied mathematical knowledge to build mathematical models and to solve these models by using the software MATLAB. Finally, a major mathematical modelling example is given corresponding to the mathematical knowledge of this experiment. After the class, students carry out the experiments needed to build mathematical models and solve the models with software for a practical problem. Students are divided into groups, they discuss the problem and finish the experiment tasks.

"Engineering Mathematical Experiment" requires students to submit 2 or 3 reports for comprehensive experiments each semester. "Mathematical Modelling Experiment" requires students to submit 3 reports.

3. Examples

In this part, we will give examples used in our course of Mathematical Experiment.

After teaching the Taylor's Expansion Formula:

$$f(x) = f(x_0) + f'(x_0)(x - x_0) + \frac{f''(x_0)}{2!}(x - x_0)^2 + \cdots + \frac{f^{(n)}(x_0)}{n!}(x - x_0)^n$$
$$+ R_n(x)$$

in the course of "Engineering Mathematics", in order to let students understand the idea of function approximation, we arrange a corresponding experiment in "Engineering Mathematical Experiment". In the class, we use the software MATHEMATICA to study the Taylor's Expansion Formula for the function $y=\sin x$. At the point $x_0=0$, we respectively compute the 1st, 3rd, 5th, 7th and 9th order expansions of $\sin x$ and draw the curves of these functions and $y=\sin x$ in one figure, using different coloured lines. When limiting x to [-0.5 0.5], we get Figure 1.

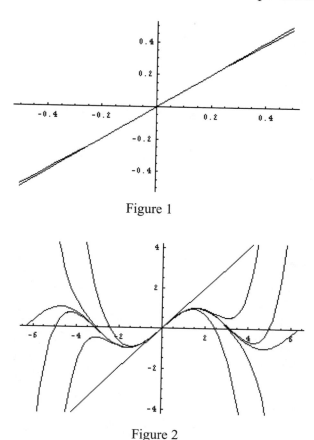

Figure 1

Figure 2

Figure 2 shows the situation when x is limited in $[-2\pi, 2\pi]$.

Then we ask students to discuss and draw a conclusion about the approximation features of Taylor's Expansion Formula from these figures. At the end, we give experimental tasks for students to do after class:

(1) At $x_0 = \dfrac{\pi}{2}$ and $x_0 = 5$, respectively compute the 5th order Taylor's expansion polynomials of sin$x$;

(2) Draw the curves of these functions and $y = \sin x$ in one figure, using different coloured lines;

(3) From the figure you get, discuss the meaning of using Taylor's Expansion Formula to approach a function.

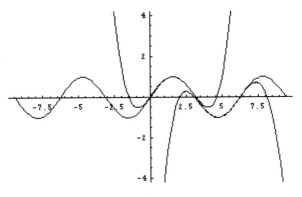

Figure 3

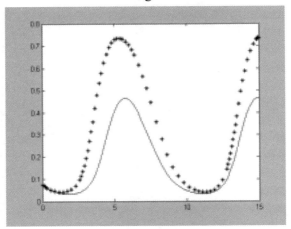

Figure4

After class, most of the students could get Figure 3 themselves and obtain the right meaning of using Taylor's Expanding Formula to approach a function.

This experiment is designed to help students to understand the theory of "Engineering Mathematics" more deeply.

In the students' attempts to solve real world problems, one of the difficulties is that they have not studied enough mathematics to solve mathematical models. For example, the students could get numerical solutions to differential equations only after they have studied the course of Numerical Analysis. Now that we have the course of "Mathematical Modelling Experiment" to help students overcome this difficulty using a computer, the students could focus their attention on building mathematical models.

For example, in one class of "Mathematical Modelling Experiment", we present the Mediterranean Sharks Problem : *From the information about the percentage of the fishing catch in the harbours of the Mediterranean, Italian biologist D'Ancona found that the percentage of sharks grew during the World War I.* Our task is to build mathematical models and give an explanation to this phenomenon.

Because the students already have a good foundation in calculus, they could set up a system of differential equations for this problem following our guide.

$$\begin{cases} \dfrac{dx_1}{dt} = x_1[(r_1 - e) - \lambda_1 x_2] \\ \dfrac{dx_2}{dt} = x_2[-(r_2 + e) + \lambda_2 x_1] \end{cases}$$

Here, $x_1(t)$ is the quantity of sharks' bait, namely the fish, $x_2(t)$ is the quantity of sharks; the growth rate of fish is r_1, the death rate of sharks is r_2; λ_1 represents the ability of sharks to catch fish, λ_2 represents the enhanced growth of sharks due to their ability to catch fish; e is the effect due to the efforts of the fishermen.

Suppose $r_1, r_2, \lambda_1, \lambda_2, x_1(0), x_2(0)$ are the same before the war and during the war. And we give the data as following:

$$r_1 = 1, \lambda_1 = 0.1, r_2 = 0.5, \lambda_2 = 0.02, x_1(0) = 25, x_2(0) = 2.$$

During the war, fishermen decreased their activity comparing with that before the war. Suppose $e = 0.3$ before the war and it decreased to 0.1 during the war.

Now we get two systems of differential equations respectively describing the situation before the war and during the war as follows:

$$\begin{cases} \dfrac{dx_1}{dt} = x_1(0.7 - 0.1x_2) \\ \dfrac{dx_2}{dt} = x_2(-0.8 + 0.02x_2) \\ x_1(0) = 25, \ x_2(0) = 2 \end{cases} \qquad \begin{cases} \dfrac{dx_1}{dt} = x_1(0.9 - 0.1x_2) \\ \dfrac{dx_2}{dt} = x_2(-0.6 + 0.02x_1) \\ x_1(0) = 25, \ x_2(0) = 2 \end{cases}$$

After that, with the help of MATLAB, we get the numerical solutions of the two equations above and draw the curves of the percentage of sharks as in Figure 4. In Figure 4, the '*' represents the percentage of sharks during the war, the line represents the percentage of sharks before the war. So we come to the same conclusion as D'Ancona.

4. The Effects of the Course of Mathematical Experiment

Through the research and practice for the course of mathematical experiment, we have developed students' ability of "Applying Mathematics", as well as trained our teachers and pushed forward the education reform for engineering mathematics of our university.

Through mathematical experiment, students' interest in and knowledge of mathematics have increased. For example, after teaching linear algebra, we gave the experiment "Schmidt Orthogonalization, Geometry Transformation, Characteristic Vectors of Linear Transformation", the students were interested in it very much. One student wrote: "I never thought before those marvelous curves are drawn by mathematics, especially by linear transformation we are studying. How wonderful! This arouses my enthusiasm to mathematics again and I decide to study it well." Students not only mastered the usage of the mathematical software in common use, but also had the idea and initial ability to analyze and solve problems using mathematical thought. Students' mathematical quality and creative ability have increased.

Through giving the course of mathematical experiment, compiling textbooks and CAI software, teachers' ability levels have increased. On the one hand, the teachers proficiently mastered the usage of the mathematical software, then used in our scientific research work and made progress. On the other hand, we mastered the method to work with CAI software and came to an understanding of the new teaching method of CAI teaching.

Through giving the course of mathematical experiment, we made some progress in course design, textbook writing, reform in teaching materials and teaching methods for engineering mathematics.

Reference

Guo Xibo, Xu Xiaoge, and Wu Guangxu (1999) 'Mathematical Experiment Course is an Important Way for Mathematical Education Reformation', *Journal of Mathematics for Technology*, 15, 60-64

20

Teaching Patterns of Mathematical Application and Modelling in High School

Tang Anhua, Sui Lili and Wang Xiaodan
Teaching & Research Group of Mathematics, Beijing No.15 High School, Beijing, China
sll180@sina.com

Abstract

The theoretical basis and teaching principles of the teaching patterns of mathematical application and modelling in middle school are discussed. Also the functional aims of five teaching patterns, teaching procedures and teacher-student functions are fully discussed in the paper.

1. Introduction

A teaching pattern is a tactical system of relatively stable teaching procedures and methods. It is formed on the basis of teaching ideas and laws and should be followed in the teaching process. Since mathematical application and modelling plays an important part in cultivating students' enterprising spirit and practical ability, it is a great change from the traditional mathematical teaching to develop mathematical application and modelling both inside and outside class in middle school. The change will inevitably bring variety to the teaching method as well as the organizational forms, and finally will break through the traditional teaching patterns and result in producing a new one. Although there are various teaching forms in mathematical application and modelling which are different in style and characteristics, they can be divided, we think, into five patterns, Pattern of Cut-in, Pattern of Special Subject, Pattern of Investigation Report, Pattern of Paper Discussion and Pattern of Mini

Scientific Research.

2. Theoretical basis of teaching patterns of mathematical application and modelling

2.1 Integration of Theory with Practice

We should persist in the principle of integrating theory with applications when studying mathematics. Through the teaching activity students know that mathematics comes from applications and can be found everywhere; they also know that mathematics is so useful that many problems in life and industry can be solved with it. Mathematical theory will show its vitality only when combined with or servicing applications. To study mathematics is to use it. Only persisting in the principle of combining theory with applications, can we truly understand and grasp this knowledge.

In the teaching of mathematical application and modelling in middle school we should provide for students an environment of learning and using mathematics. By their personal exploration, discovering, analysis, study, solution and test – the whole process of solving a problem, students get not only the experience of learning and using mathematics but also the consciousness of watching and analyzing the matters beside them with mathematical knowledge they have learned.

2.2 Outlook on Study of Constructivism

Constructivism does not believe that knowledge is simply the passive transmission from teacher to learner and regards it as an initiative process of constructiveness on the basis of a learner's knowledge and experience, that is, all the knowledge comes from constructiveness. In the process of constructiveness the existed cognitive structure of the learner plays an important role and develops continuously.

In the teaching activity of mathematical application and modelling students find out the problems by investigation; they change the problems into mathematical ones and then find the way to obtain a solution. They seek and gather materials, learn from the person who knows how when encountering difficulties, and finally solve the problems. This is a process of learning and doing mathematics, and, much more, is a process of initiatively constructing cognitive structure.

2.3 Key Issue to Education Reform is Student's Subjective Participation

The key issue in the present reform of promoting quality in

mathematical education is to persist in and develop the student's subjectivity and in mathematical teaching, subjective participation is an important principle consistent with the above issue. In the teaching patterns of mathematical application and modelling, especially in Pattern of Special Subject, Pattern of Paper Discussion and Pattern of Mini Scientific Research, on the basis of teacher-student's equal participation the problems are solved not according to the teacher's methods and procedures designed in advance, but the student's thinking, methods and procedures. This kind of activity provides an opportunity for students to explore, to put forward and to solve problems by themselves, which totally show the students' subjectivity and the principle of students' joining teaching activity. It makes the students learn in a process of learning, to explore in a process of exploration. It also trains the students' ability to solve problems in practice.

2.4 Innovation Education is the Heart of Quality Education.

It was emphasized by President Jiang Zemin of China that innovation is the soul of a nation and is the continuous driving force to make a country flourish. The crux of innovation is qualified people, that is, there must be many highly qualified people with innovating sprit. Training qualified people with innovating sprit requires social engineering and should be made among children at different educational levels. The innovative education at the stage of elementary education should address the whole student, lay stress on cultivating student's innovation consciousness and sensibility and lay a foundation to form an innovative nature and to train innovative ability. The essence of quality education, which we advocate today, is to develop proficiency in each student and to provide the education that each student needs for his or her future development.

However from ancient to modern times the experience of the scientists who made a great contribution to the process of humankind reveals a truth that it is not enough to pass on knowledge in school education. The purpose of passing on the essence coming from predecessors' experience and knowledge to students is to make them stand firmly on the predecessors' shoulders and to create a brighter future. So the key point of education should be arousing and leading the student's exploration and innovation. Teaching of mathematical application and modelling in middle school requires learning and grasping the theoretical scientific law as a prerequisite. It uses mathematical knowledge already learned as the basis and makes students learn mathematics in a process of scientific exploration and research, applying mathematics in nature, which is the best way to promote a student's comprehensive quality and to develop their latent creative power.

3. Teaching Principles of Mathematical Application and Modelling

The application of mathematical knowledge includes two kinds of application, the application inside mathematics and outside it. The teaching of mathematical applications mainly considers examples with a practical background. The teaching aim to be reached is to understand more deeply the knowledge and methods that are already learned, and to reinforce them. Another aim of teaching is to address the attitudinal problem that mathematics is useful. Mathematical modelling uses mathematics to solve problems that are non-mathematical in life and industry. Although limited by knowledge and ability, most problems about mathematical modelling in middle school still have the nature of application. They still need to gather data, construct a mathematical model, solve the model and test it in practice. So the following teaching principles should be reflected in the teaching patterns of mathematical application and modelling.

3.1 Principle of Recreation

The class teaching of mathematical application and modelling provides students with the possibility and opportunity to put forward problems and to solve them by themselves. So the key issue of mathematical modelling is the activity of recreation under the prerequisite of students' active participation.

3.2 Principle of Mathematization

Students draw problems from applications and change them to problems of pure mathematics, that is, they learn mathematics in the process of mathematicising practical problems. What we valued is to help students to think in mathematics and to observe the world in mathematics terms. So the teaching process confirms the well-known saying from the famous Netherlands mathematician, Freudental: "It is better for us to say learning mathematization than learning mathematics".

3.3 Principle of Mathematical Feasibility

In our teaching we fully recognize and emphasize a student's personal characteristics. We launch mathematical application and modelling of different levels among students with different ability and provide, as soon as possible, the different environments to different students to show their creative ability. All students can use mathematics in varying degrees and can get the experience of mathematical application with different levels in the process of using mathematics. On the basis of his mathematical reality each student will heighten his mathematical ability, application consciousness and practical ability.

3.4 Principle of Rigorousness

As for mathematical modelling in middle school it is unnecessary to pursue sedulously great complexity and perfection in the modelling process. It is also unnecessary to apply absolute rigor in simplifying the model and calculating the answer. It should be the rigour within the conditions of a student's mathematical reality. So there should be a special evaluating standard for the result of students' modelling.

4. Teaching Patterns of mathematical application and modelling
4.1 Pattern of Cut-in

It is a teaching pattern of cutting mathematical knowledge in some links of the whole teaching process and is not necessary to contain the whole process of solving a problem.

Functional Aim

In class teaching the teacher should introduce applied mathematical examples to arouse student's interest, and to let them see the use of related mathematical knowledge and learn that mathematics is useful. Another aim is to encourage students initially to grasp the way of solving application problems with related mathematical knowledge.

Teaching Procedures and Teacher-student Interaction

At the beginning of teaching new knowledge, a new concept or a new theorem the teacher provides the historical data and introduces some application examples with practical background. He or she creates an application environment to draw students' attention and arouse their curiosity as well as satisfy their thirst for knowledge and exploration. At the stage of strengthening the knowledge, such as reinforcing concepts and theorems, the teacher introduces the processed practical problems. Now the key point of the teaching is to strengthen the learned knowledge. e.g. At the beginning of the lecture about sequences of number, the story of the Indian King planning to reward highly the inventor of chess with wheat was introduced. After learning geometric progression, the students would deal with such problem as how to calculate the interest rate of saving account in the banks.

The teacher creates an application environment for the students. Teaching is active while the students are passive. Lead by the teachers the student, with active thinking, can finish the process of changing the problems into mathematical ones and solving them. It shows that students are the subjects in the teaching process.

Evaluation

The application problems in Pattern of Cut-in, in general are processed and have the condition that is clear, accurate and sufficient. The knowledge used for solving these problems is single and clear. The process to transform the original problems to the one of mathematics is also simple and clear. The conclusion is single and hardly the consideration whether it meets the application or not is needed. So these application problems are simple and special modelling problems. They do not show well enough the typical process of mathematical modelling and cannot reflect the creative and open nature of modelling. Therefore Pattern of Cut-in has the strong function of cultivating students' application consciousness and making them grasp the way of solving problems with mathematics. But the function is limited in training students' practical ability.

4.2 Pattern of Special Subject

This is the pattern that after teaching one or several chapters, or encountering the cross connection of knowledge, the teacher, together with students, solves the application problems.

Functional Aim

Concentrating time and energy, the teacher teaches mathematical knowledge and directs students' thoughts to lie in an application environment in the whole teaching process to deepen their application consciousness. By strengthening the knowledge already learned the teacher makes students appreciate the way of mathematical application and of solving practical problems with mathematics. By the means of modern educational technology, such as calculator and computer, the aim of training the students' ability to experiment mathematically is achieved.

Teaching Procedures and Teacher-student Interaction

The teaching aim of Pattern of Special Subject is, in a sense, a review and summary. Carefully choosing the problems is a prerequisite to give a better lesson of mathematical application in Pattern of Special Subject. The principle of choosing examples is to pick them up carefully and determine the key point of teaching progress according to mathematical knowledge, e.g., the application of function, the application of number sequence, the application problems of measurement, the application problems of finance and economy, etc. The problems dealt with in Pattern of Special Subject are more difficult and more open than the one in Pattern of Cut-in.

In the lecture on the application of the knowledge about Functions to students of Senior One, the teacher provided a set of population data collected every ten years from 1790 to 1980, and organized the students to analyze it and discuss the following questions:

a) What do you find in it?

b) How many volumes are there?

c) Observe the Point Graph. What Function Graph, which we have learned is similar to the changing trend of the points? Figure out an Elementary Functions roughly showing the changes of the plotted points, construct the graph on the computer, find out the equation and analyze the error.

d) Foresee the population of this country in the year 2000.

e) Compare the matching of the point graph and the graphs of linear function, quadratic function, exponential function and power function.

Finally the teacher concluded the operating process of solving such problems and the thinking process of dealing with practical problems and assigned the students to foretell the population in China as homework. This case best illustrates the "pattern of Special Subject". The examples carefully chosen by teachers provide only a good question environment for students. In teaching process it gives no longer the first place to teacher's lecture. The teacher gives carefully chosen examples to students. After reading the examples carefully, analyzing the data and information and looking for the deeper information students put forward a model hypothesis and solve it. Students finish the whole process of solving the problems independently and cooperatively. The teacher, a provider of the problems, functions as a modelling advisor and an organizer for class discussion and evaluation, while students are the vital new force to solve problems and the subject of class teaching. It realizes the transformation from teacher's 'teach' to students' 'do' and demonstrates the outlook on study that learning mathematics is achieving.

Evaluation

The student's main task in class is to solve the problems offered by teacher using existing knowledge. Most examples are processed. The background analysis and the process of mathematization to practical problems become the key point of the teaching. Although the mathematical model is need, it cannot show totally the typical process of mathematical modelling. Such modelling is not completed, which cannot demonstrate fully the creativeness and the openness in mathematical modelling. Although Pattern of Special Subject is better than Pattern of Cut-in in training students' application consciousness, enterprising sprit, practical ability, etc., it still has the limitation.

4.3 Pattern of Investigation Report

This is a new teaching pattern combining the class teaching with activity outside class and is a pattern of investigation outside class and exchange inside class.

Functional Aim

With a process of social investigation, data collection and the discovery of problems outside class, students learn how to watch and think about the world with mathematics. They strengthen their consciousness of solving practical problems with previously learned mathematical knowledge and they develop their sense of curiosity and their ability to find problems. Because the problems may be different from one other, the methods of modelling and solution are various. They are flexible and gradational. Using computer to choose, process and arrange the data, students experience the way of data processing and develop the skill of data processing. Pattern of Investigation Report creates an environment for close cooperation, division of labour and brainstorming for students, and meanwhile trains their cooperation sprit.

Social investigation makes students go outside the school and puts them in a learning environment that is dynamic, open, positive and multi-faceted. It provides for students more channels and ways to get knowledge and cultivates their sense of responsibility to society and life. It heightens their awareness and their ability to interact socially.

Teaching Procedures and Teacher-student Interaction

First the teacher should explain the aim and objective, that is, knowing how to apply some mathematical knowledge in life and finding out which mathematical knowledge is useful in a certain field. After learning the formulae for the Surface Area and the Volume of examples of

Prism, Pyramid, Frustum and Sphere shapes, the Senior One students were assigned to find out ways of packing drinks like milk and Coca cola. They were to analyze and discuss how much material is required in the different ways and to figure out which packing design benefits the drink company most and which benefits customers most. Students make an outline for investigation in groups and gather the data in cooperation with a due division of labour. They analysis, extract, process and transform the data to get the mathematical model. They solve the problems with collection and discussion, and they write investigation reports. For students who are lower in ability or for whom the problems are more difficult, it is necessary for the teacher to discuss with them the modelling and solution methods. Finally some time should be taken in the whole class to discuss and evaluate the project under teacher's guidance and organization. At the same time the teacher gives a brief and well-prepared summary. Besides the full affirmation of achievements, it is necessary for the teacher to point out the general problems and deficiencies in organization, technology and modelling.

It is obvious that in the teaching process of Pattern of Investigation Report, the teacher is an organizer and commander of the activity; meanwhile he is a cooperator with students' investigation and modelling. Students launch the social investigation, put forward the problems, make mathematical models and solve them themselves. They are no longer people who learn knowledge passively but become people who advocate and solve problems. They initiate new cognitive structures in themselves.

Evaluation

The students solved truly practical problems. The pattern demonstrates the whole process of mathematical application and modelling. Compared with Pattern of Cut-in or Pattern of Special Subject, it is more efficient for training students' application consciousness, enterprising spirit and practical ability. Since the teacher assigns the content and framework of the theme of investigation and research to be done by students, all the work students do is to realize the teacher's teaching purpose. This places a restriction to students' thought and limits the openness and educational function of the pattern. Similar to Pattern of Special Subject, Pattern of Investigation Report is suitable only at a certain teaching stage.

4.4 Pattern of Paper Discussion

In Pattern of Paper Discussion the teacher chooses the papers carefully, students read them outside class and discuss them in class led by one of the students. It is a pattern to learn mathematical application and modelling by means of studying and discussing other's excellent modelling papers.

Functional Aim

The pattern asks each student to read carefully the modelling papers following the teacher's demand. The aim is to train students' independent learning ability and, especially, to make students notice the difference between mathematical papers and other articles. The students can receive a training of studying mathematical papers, which will lay a foundation to students' life-long study in their learning behaviour and method. By reading alone and discussing together students should try to understand the papers in the following different points:

a) Knowing what problems the papers discussed and what their background is. Knowing papers' structure and main point.

b) Knowing papers' modelling methods, steps, tactics, process and conclusions.

c) Feeling author's modelling thought and method.

d) Giving their own opinions to the thesis and giving their own modelling thought.

Although learning another's mathematical modelling is different from making a model by oneself, learning how to make a model from another's modelling thought and methods is doubtless meaningful in training students' independent study ability. Since the chosen papers are from middle school students and readers' knowledge and thinking style are similar to the one of authors', this teaching pattern can realize the aim of highly promoting a group of students' abilities in mathematical application and modelling with less time and better topics.

The question circumstance of paper discussion provides a wide field for students' enterprising thought. It also provides a chance to show an original view coming from students' thought. Different to the usual class teaching, Pattern of Paper Discussion pays more attention to students' initiative in constructiveness. The equal participation of teacher and students creates a studying atmosphere that is democratic, harmonious and comfortable. It makes the students more independent.

Teaching Procedures and Teacher-student Interaction

In the teaching procedure of Pattern of Paper Discussion the teacher carefully chooses a paper of mathematical modelling about 10 days before the lesson – a paper from middle school students is more suitable. The teacher prints and distributes the paper to students, and at the same time assigns the reading requests that asks students to study the paper's problem background, modelling methods, solution tactics, solution methods, etc. Students read the paper outside class and write out reading notes with ideas on how to construct the model if he encounters this practical problem. One or two classmates are chosen as the host of class discussion. In the discussing lesson the host first states his opinions about the paper's background, structure and conclusion. He also discusses the paper's modelling process and modelling methods. Then the classmates put forward the questions about their understanding and the host's explanation. The host answers these questions or organizes students to discuss. Teacher and students participate in the discussion equally to express their opinions and to put forward questions.

The following two points mainly show the teacher's leading function.

a) Carefully choosing the papers, creating a good questioning environment, pointing out the reading links and putting forward the reading requests to arouse students' activity and creativeness of independent study and exploration

b) The apt summary to the whole discussing process

Teacher and students participate in the discussion equally. Both of them are the reader and researcher of the paper. The teacher is no longer the spirit of correctness. Neither is he or she a play writer, a director or a leading role. Instead he or she is the students' learning and discussing partner. Sometimes the teacher gives some advice to the host and acts as an advisor; for most of the time he or she is a loyal listener to the students' host and a proponent or opponent of the argument. The students are the speakers of paper discussion and the initiating constructor of their own cognitive structure.

Evaluation

In Pattern of Paper Discussion students see that mathematical application problems are no longer general. The background of these problems may be a non-mathematical field and the way used is the combination of learning outside class and discussing inside class. It breaks through the traditional pattern with which students learn knowledge by means of teacher's teaching and is more active, flexible and open. But it is such that make the following points are most important.

) Selection of papers

) Selection of host

) Understanding paper's modelling thought, solution method and solution tactics.

4.5 Pattern of Mini Scientific Research

This is a teaching pattern that students choose and determine the research subject independently from life or industry. Then they carry out scientific research on a small scale and write their research results in a paper. The report and public presentation can train students' consciousness of mathematical application, enterprising spirit and practical ability. It is a pattern that stresses personal or group activity.

Functional Aim

The students go deep in society without any restriction in their thought. They choose and determine research subjects they are interested in from daily life. In the modelling papers they collectively state their research methods, steps, views, results and their cooperative experience. They also organize the report and public presentation. This kind of activity not only changes students' learning place and method but also emphasizes independence and exploration in learning process. It demonstrates more the practice, openness, independence and process in study. So it has a stronger function in cultivating students' innovative consciousness, enterprising spirit, practical ability, even their general quality and ability.

Teaching Procedures and Teacher-student Interaction

The Pattern of Mini Scientific Research can realize study with research characteristics. After students determine their research subjects they must do a lot of investigation on the historical and present condition of the subjects. They carry out a mini research project according to the subjects. From choosing the subject to writing the paper, in the whole process students actively make each decision and do each job according to what is required. They put forward the subjects, design the scheme and carry out it. The teacher should mainly guide the work.

First-phase Guidance

Before students begin to select the subjects they receive guidance collectively from the teacher. The guidance is useful in the following three aspects.

a) With a lot of examples teacher shows that mathematics is important to human civilization, the development of high technology in the future, the competition of national strength and the growth of qualified people. The papers written by middle school students demonstrate the fact that middle school students can participate in scientific research on a small scale and write their research results in a paper.

b) The teacher tells students how to choose a subject. With some practical examples, especially the one at hand that seems irrelevant to mathematics but actually contains some mathematical principles, such as the problem of people evacuation, the number of serving windows in a dining room, etc, the teacher demonstrates that mathematics is used extensively in daily life, in industry and work. These examples not only enlarge students' views but also make them know that only by paying close attention to society and life, can they find subjects worthy to study. Because middle school students have little learning and have insufficient research methods, which is the subjective condition, and the problems may have so many factors that we cannot forecast them in advance or they involve too many things, which is the objective condition, it is possible that the students could not research all the subjects they found. We think that students should choose the subjects with which they are familiar and interested. Also they should choose the ones that can demonstrate their strong ability and that have fewer variables.

c) The teacher should guide the students in the methods of scientific research.

Mid-term Guidance

Students make a collective exchange at a given time and receive the teacher's guide respectively. Each student or each group of students – we advocate that two students cooperate and finish the paper – introduces his chosen subject and the work he has finished or is doing. Then the teacher makes exchange and gives guidance to each student or each group of students.

Later stage Guidance

After the teacher reads each paper, he questions each paper and gives an opinion to revise it. The teacher also asks the students to test the results expressed in their papers in practice as far as possible and to make a proper analysis of the testing result. Sometimes the teacher guides a student for his papers more than one time. The guidance does not stop until the paper is finalized.

When the papers are finalized, the guiding teacher should read them one by one carefully. The teacher selects some papers and organizes a public reply. On the public reply other teachers of the same subject and experts appraise these papers. A public presentation of papers is held among the students of the same grade. From each lecture more students can get the education of independent study, mathematical application and enterprising spirit; meanwhile they are aroused and inspired by their contemporaries. They believe that mathematics is useful. Also they are confident that they can write papers, find out the problems beside them and solve these practical problems with their mathematical knowledge. At the same time they will understand the meaning of knowing one's deficiencies after learning and arousing a stronger thirst for study. They will learn mathematics harder to lay a solid foundation.

As the last step of the teaching process, a symposium is held in which the teacher and students discuss what they have learnt from the research and study.

In the teaching process of Pattern of Mini Scientific Research the teacher has no intention of interfering in the students' choosing research subjects, the students' modelling thought and their researching methods. It is also impossible to replace what the students should do by teacher's work. So the teacher is a teaching organizer, an advisor to students' research and an expert judge of students' papers.

Evaluation

Pattern of Mini Scientific Research creates a most open innovative environment. The first issue to write an excellent modelling paper is to choose the subject and a good research subject comes from an author's sharp and careful observation of the outside world. The enhancement of the insight depends on active observation and the cultivated behaviour of being ready to observe. It also depends on keeping one's curiosity as well as having a spirit of analyzing and criticizing the existing conclusions. The behaviour, ability and spirit are the basic qualities of an innovative person. Students will encounter many difficulties and setbacks in the whole process and even be defeated, if they made a wrong decision. However it is the stubborn and unyielding courage, confidence and willpower to overcome the difficulties and setbacks that an innovative person must have.

Summary

The aim and function of mathematical application and modelling ask us to appreciate the development of students' innovative consciousness, practical ability and general quality. We should fully confirm what the students selected and what they did in their research. We should not think too much about the application value of what they did and should not emphasize too much that the problem be new or that the method be original. It is very good for a middle school student if he can discover rules new to him but known to others by his continuous efforts and dependable works. If students can study practical problems with the way of mathematical modelling and get some results, we should say this is the evidence of their innovative consciousness and ability. Facts have proved that the significance of students' writing papers on mathematical modelling goes far beyond the meaning of the problems discussed in their papers. This is why the Pattern of Mini Scientific Research is valuable.

5. Conclusions

a)   Teaching forms of mathematical application and modelling are various and can be summed up in five patterns. These patterns have different teaching aims and are suitable to different stages in the teaching process.

b)   The above five teaching patterns have the same theory basis and teaching principles. But they have different emphases on the aim of training students' enterprising spirit and practical ability and show the different levels, high or low. Among them Pattern of Mini Scientific Research is the highest level of mathematical application and modelling in middle school.

c)   The first four patterns must be the basis for Pattern of Mini Scientific Research. Trained by the first four patterns, the students have an initial experience of mathematical application. They receive practical training in Pattern of Investigation Report and prepare for mathematical modelling and paper writing with Pattern of Discussion. Together with some basic mathematical knowledge they can launch smoothly Patterns of Mini Scientific Research and get the expected result.

d)   The excellent modelling papers from students are a valuable resource for teaching in the future. They should be arranged, be bound into book form and be added to continuously.

Teaching pattern is the reflection of teaching theories, ideas and laws. It is also the integrated system of the teaching methods, procedures and organizational forms. The above analysis and research provide a system that is useful as a reference and is easy to operate in the teaching of mathematical application and modelling in middle school. However a pattern should not be the new restriction to our teaching and the teaching of mathematical modelling should not be stereotyped. Furthermore a pattern should not be used just as a mere formality in order to chase its integrity. To study the patterns is to jump out of them. With the continuous enrichment and accumulation of the practical experience in a teacher's modelling teaching and the continuous revision and renovation of educational theory, teaching ideas and teaching concepts, the original teaching patterns must be broken through. It will produce such a result that our teaching of mathematical application and modelling will become the one with a teacher's personal style and characteristics. As an open system such a teaching pattern is vital. It has the function of promoting the benign development of mathematical teaching and the function of promoting the reform of mathematical education in middle school. So the study on mathematical application and modelling in middle school has not finished at present and will not finish in the future. There are still many problems to be discovered and researched. Our task is heavy and our road long. What we should do is to try harder.

Bibliography
Yan, S-J (1996), 'Mathematics education in 21 century' (in Chinese), Nanjing, China: Jiangsu Education Press.
Ye, Q-X (1998), 'Mathematical modeling for high schools' (in Chinese), Changsha, China: Hunan Education Press.
Zheng, Y-X (1999), 'Modern development of mathematics education' (in Chinese), Nanjing, China: Jiangsu Education Press.

21

Mathematical Experiment Course: Teaching mode and its Practice

Qiongsun Liu, Shanqiang Ren, Li Fu and Qu Gong,
Chongqing University, Chongqing, 400044, P.R.China
liuqiongsun@163.com

Abstract

In this paper, we explain the role of mathematical experiment course in the education reformation in China. We also describe the objective of the course and the students who should take it. The content and framework of mathematical experiment course are established and optimized. We explore the teaching mode and approaches for mathematical experiment course. We summarize all the ideas in teaching the course and in mathematical course reformation carried out in Chongqing University.

1. Introduction

In the mathematics education of undergraduates, one of the most important issues is how to reinforce training the ability of applying mathematics and how to help student to master the skills to solve real problems with the aid of a computer. This is an important in college mathematics education reform all over the world. It is agreed by education experts, both domestic and overseas, that mathematical experiment course is a solution to the above issue (Fengshan Bai et al, 1997).

Mathematical experiment is a new idea which came from the availability of computer technologies, especially mathematical software. It tries to reform the teaching infrastructure for mathematics education, including teaching content and methodology. Mathematical experiment offers new content and course type for mathematics education. The traditional mathematics education infrastructure has emphasized students' capabilities in logic, geometry and calculation. When a student confronted a real problem and tried to describe it with mathematical language, i.e., model it, then solve it and verify the result, the traditional way turned out to be insufficient.

We think that three capabilities should be stressed in college mathematics education:

1. Consciousness of applying mathematics to describe and solve practical problems;
2. Abilities in mathematical abstraction and mathematical modelling;
3. Abilities in mathematical analysis, computing and in using mathematical software.

In recent years, many colleges and universities have started using mathematical experiment course. It has been emphasized particularly on modelling approaches and oriented to MCM (Mathematical Contest in Modelling). This has contributed to training the student capabilities mentioned above. However, mathematical modelling deals with so many mathematical branches, so students have to spend a lot of time in preparing background knowledge. This is difficult especially for freshmen and sophomores. Mathematical experiment does not emphasize mathematical knowledge itself, rather the "experiment". By doing the experiments, which mostly originate from practice, students can master the whole procedure to apply mathematics. The procedure typically consists of mapping the real problem to an appropriate mathematical problem and applying mathematical theory and methods to solve the problem effectively, with the aid of a computer. Mathematical experiment emphasizes particularly a 'do-it-yourself' approach to students, really trying and experimenting with things. This will effectively improve students' abilities in initiative and practical operations (Pollatsek, 1998).

2. Course content and teaching mode

2.1 Outlined design of the course content

First of all, we must make clear the aim of the mathematical experiment course. Secondly, we must know who should take this course. We need to establish the infrastructure of the teaching system and optimize the structure. We have designed the modular form of teaching content: four modules are recognized according to modern mathematical science, i.e. numerical computing, probability theory and statistics, optimization techniques and graph theory and network optimization. Each module is divided into sub-modules (Fu Li et al, 2000), see Figure 1. The software used is MATLAB.

In the design of teaching content of mathematical experiment, we have emphasized the relationship between different courseware and the versatility of approaches and methods to solve practical problems. Training students is arranged in several stages according to the step-by-step principle. In the detailed design of the course content, both the relationships between modules and the independence and completeness of each module are considered. Specific requirements are made for each training stage, so that the students can understand the basic ideas and concepts, methods and algorithms, and the whole process to solve a real problem. This will make the students master the skills required to do research and to develop independently. Experiments fall into two main parts: fundamental experiments and comprehensive ones. For example, forest management model, lake pollution problem, DNA serial classified model based on statistical method and neural network, equipment update problem, etc.

2.2 Teaching methods and mode for the mathematical experiment course

The teaching mode is one that emphasizes students' independent operation with help as necessary from teachers. The teachers' tutorial should be in a heuristic manner. We present an interactive students-teachers teaching mode. The traditional teaching mode is teacher-centric with one teacher to many students. This situation has not changed for a very long time. The new multi-students multi-teachers teaching mode is essentially students-centric: more than one teacher teaches the course. This teaching mode is largely discussion based.

The new teaching approach has advantages in students' active learning supervised by teachers. The students cooperate to accomplish tasks. The observed "group effect" will improve the total teaching achievement. The cooperation between teachers makes best use of all the teachers. The students are encouraged to be thoughtful and creative. Students are allowed to choose among experiment tasks to do modelling, programming and result analysis. Our teaching programme introduces a feedback mechanism. The teaching plan, including the corresponding teaching content and materials, is regulated according to instant, mid-term or long-term feedback on information from students to optimise teaching effects. So any teaching plan is implemented in a closed-loop manner rather than an open-loop one. This also reflects the idea of student-centric but not traditional teacher-centric teaching. We designed the content to show this pattern: "introducing example-summary of concepts and methods-software-samples-practice". Theoretically, this subsumes all the processes: from concrete to abstract and then back to concrete, from specific to general then back to specific, and from problem to methods and then back to problem. It also subsumes induction and deduction, analysis and synthesis. When we introduced the new teaching mode in the mathematical experiment course, the students were very interested. They felt the charm of mathematics and its application in engineering and science.

2.3 Multimedia courseware research and development for mathematical experiment

As an important part of modern educational technology, CAI (Computer Aided Instruction) has been developing in China (Weizhu Huang et al, 1999). It plays a very important role in promoting education reform. Since 1999, we have explored many forms of multimedia courseware for mathematical experiment course; we put it into teaching practice and achieved much.

We authored the textbook for mathematical experiment (Fu Li et al, 2000). In this book, we introduced a "show-up model" for the teaching content: introducing example – knowledge and concepts summary – related MATLAB functions – samples – experiments. And we designed the style of the book by presenting "think", "do", "tips", and "remarks" icons. This is one of the most important features of the textbook. We also developed the corresponding electronic form of this book in PowerPoint. Considering that students' demands will change, the teaching system should be scalable and expandable. Besides, different teaching styles of teachers and different choices of students have been considered as well. For the above reason, we take three main approaches in organizing teaching:

1) Using PowerPoint. The background of the practical problem, related mathematical theory and methods, related MATLAB functions, and samples are showed on the screen. A table of content is designed into a tree structure; each knowledge leaf can be reached along this tree. After the teacher completes one "leaf", he/she can go back to the root of the tree, and then can go anywhere of the tree structure. The modification and complement of the tree is very convenient. This approach is very similar

to, or essentially the same as hyperlink text used in web pages. Frame-figure for teaching content is designed as so:

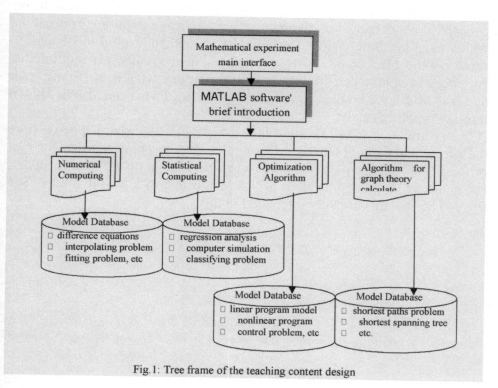

Fig.1: Tree frame of the teaching content design

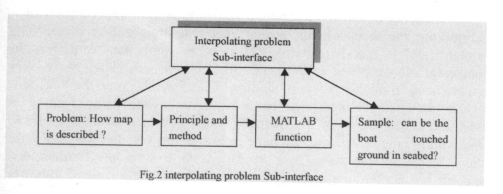

Fig.2 interpolating problem Sub-interface

2) We use the mathematical software tool to implement numerical computing, symbolic operation and graphical displaying. It can jump from a PowerPoint page to MATLAB. Some teaching content is related to the concept of convergence or divergence in numerical computing or iteration, which even generate chaos. This kind of phenomenon used to be hard to show on the blackboard. Now it can be shown vividly via multimedia. Another example: when talking about curve fitting, what is the fitting error and is it within the bound required in engineering? This can also be shown through graphics easily.

Animation is used so that the students can see what happens. Take the predator-prey problem as an example; we can show a video for real process. We can also play simulated animation based on a dynamic model, for example, a guided missile followed the enemy's vessel tracks. See Fig.3. The students are very interested in this.

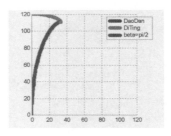

Figure 3 Kinetic track's graph

3. Reinforce the teacher team

In order to fulfill a teaching task fruitfully, one important factor is reinforcing the teacher team beside excellent organization of teaching content and a good textbook. Two years ago, the team for the mathematical experiment consisted of four teachers. They were very experienced in mathematical modelling. They have become the backbone of the mathematical experiment team. However, a four-teacher team is far from enough if we want the mathematical experiment course to spread wider and wider. Therefore, we have been training younger teachers since then. Seven people have been recruited up to present. Furthermore, specific teacher's training class was organized on behalf of the Committee of Mathematical Modelling of Chongqing. This helps reinforce the teacher team in a larger scale. This year, we made an investigation for all the colleges in Chongqing University to find that they appreciated the mathematical experiment course and the mode in which we carried out teaching activities. Some colleges even expressed the view that they would take the course as the normal teaching program for their students (Shanqing Ren et al, 2001). This impact requires more teachers.

4. Teaching effect

Since 1998 when we first introduced the mathematical experiment course, more than one thousand students have taken it. A sampling investigation on students who have taken the course shows that the course is widely accepted and welcome by them. Most of them said they "like it most" or "like it very much". Some of them said, "The teachers introduced a lot of sample problems which were very interesting. We saw many things that we had never seen before. We learned some mathematical software tools that were very powerful. We would remember the mathematical experiment course for life!"

Students are not the only people who benefit from mathematical experiment course. Teachers have also been trained during the teaching activity. They have contacted many new mathematical ideas and approaches other than traditional ones. They have mastered many new computing skills and simulation techniques. It is a leap for their knowledge structure and general capacities. All of this guarantees the continuing development of the mathematical experiment course.

References

Fengshan Bai, Jianmin Jiang et al.(1997) 'Construct the CAI module for mathematical experiment and mathematical teaching and train the students for their computing skills and creativity'. The Achievements of Twenty-one Century Oriented Teaching Content and System Reform for Higher Education in Engineering: *Challenge, Exploration and Practice*, Vol.1. Higher Education Press, 351-354.

Fu Li, Gong Qu et al (2000) textbook for 'Mathematical Experiment', Beijing: Science Press.

Pollatsek, Harriet (1998) (Ed) textbook for 'Laboratoroies in Mathematical Experimentation', translated by Bai Fengsan and Cai Dayong, Springer.

Shanqing Ren, Qiongsun Liu et al (2001) 'Setup the mathematical experiment course for the students to further promote the reform of mathematics education.' Proceeding of International Seminar of Engineering Mathematics Education and its Applications. The Hong Kong Polytechnic University, 225-229.

Weizhu Huang, Wen Zhao et al (1999) 'Education information integration-the important direction for CAI development.' The Achievements of Twenty-first Century Oriented Teaching Content and System Reform for Higher Education in Engineering: *Challenge, Exploration and Practice*, Vol.2. Beijing University of Aviation and Space Press, 474-476.

22

The Mathematical Modelling – Orientated Teaching Method of Elicitation

Ruiping Hu and Shuxia Zhang
Zhenjiang Watercraft College of CPLA, Zhenjiang, Jiangsu, China
Maple_hrp@sina.com.cn

Abstract

This paper summarizes the recent course of the Chinese teaching reform in engineering and the introduction of mathematical modelling as an effective means into our teaching reform in engineering. It puts forward the mathematical modelling-orientated teaching method of elicitation, which the authors have tried out in their teaching practice of advanced mathematics, as an effective means of embedding mathematical modelling into the classroom teaching of advanced mathematics. Some questions are illustrated by examples of the use of the method.

1. Mathematical modelling and the teaching reform of Chinese engineering courses

Chinese engineering teaching has a fine tradition, that is, great emphasis is placed on the teaching of knowledge, and as a result the students have a systematic and good grasp of basic knowledge, and have high abilities in taking examinations. In the 1990s, facing the challenges of fierce competition between countries, rapid development of international science and technology and the ever-changing characteristics of Chinese society, some educators with a breadth of vision saw that there are many disadvantages in Chinese traditional engineering teaching. The most conspicuous of these is the stifling of students' individuality and thirst for knowledge and creativity. Therefore, in the middle period of the

1990s, the teaching reform was initiated by the Chinese authorities. With the experiments and experience of the last few years, a common understanding about the reform has been reached. This is reflected in the document, *"The Decision about the Deepening of Educational reform and the Overall Carrying forward of Quality Education"* Passed by the Central Committee of the Communist Party of China and the State Council. The common understanding is to carry out quality education, the key of which is to cultivate an innovative spirit and practical ability.

In China, engineering mathematics traditionally includes such required courses as advanced mathematics, linear algebra, probability theory and elective courses such as complex analysis, integration, equations of mathematical physics and special functions, algorithmic language and computational methods the key to which is advanced mathematics. Engineering mathematics is one of the most important basic courses in engineering universities and colleges, and occupies a pivotal position in engineering course teaching. In order to meet the needs of quality education and emphasize the cultivation of an innovative spirit and practical ability, what to do with engineering mathematics is a practical and critical puzzle every mathematical educator faces. For this reason, many mathematical educators deeply and continuously probe the reform of teaching aims, curriculum, teaching content, teaching materials and methods of engineering mathematics. They have made considerable headway. Among the achievements, an influential result is the mathematical modelling competition between the Chinese college students that began in the early 1990s. With great efforts over ten years, the scope of the competition has become the most extensive among students' extracurricular activities of science and technology in Chinese universities and colleges. Teaching practice indicates that mathematical modelling activity has played a very important role in the cultivation of students' overall qualities, especially the strength of their innovative spirit and practical ability. It also indicates that teaching reform has been deepened in engineering universities and colleges, as is well known in the educational field (Qixiao Ye, 2001). However, for many reasons the traditional teaching model still occupies the leading position in present-day engineering mathematics teaching. Only some universities offer courses of mathematical modelling and experiment, and only some of the students can participate in the competition in mathematical modelling. As a result, many students do not benefit from mathematical modelling teaching. How to make up for the weakness above is a question worth studying deeply. According to the authors, it is not enough just to depend on the mathematical modelling course and the modelling competition (Ruiping Hu 1996), and it is necessary to embed modelling into traditional classroom teaching, especially into the key course of engineering mathematics – advanced mathematical classroom teaching. The mathematical modelling-orientated teaching method of elicitation is the best combination of mathematical modelling teaching with elicitation method in mathematical classroom teaching. By doing so, the previous teaching plan of engineering mathematics can remain unchanged, and the classroom teaching as a key teaching means can be made best use of, which organically combines the results of teaching practice with the fine tradition of previous engineering teaching.

During our teaching practice of advanced mathematics in recent three years, the authors consciously used the mathematical modelling-orientated teaching method of elicitation in their teaching. They achieved good results. One of the most outstanding results is to stimulate the student's interest in mathematics. For example, among our college graduates who applied for postgraduate entrance examination, nearly 90 percent selected the major of military operational research. In addition to their classroom learning, they must learn by themselves mathematical knowledge. They passed the postgraduate entrance examination, and nearly 60 percent of them were enrolled.

2. The mathematical modelling–orientated teaching method of elicitation

The "teaching method of elicitation" is, in general, the totality of the style of mutual relationship and interaction in teaching activities between teachers and students. However, here it refers to the concrete teaching approach. The elicitation method of teaching includes the following processes. First, rich and varied teaching material, selected carefully by teachers, is given to students to lead them into a state of cognitive conflict. Then the teacher makes the students cognitive structure achieve a new balance through the process of "stimulating thinking" and "inducing understanding". They do this using inducing questions, situation and analogy, and discovery, induction and research. Finally students realize the inner absorption of knowledge and skill. The teaching practice has alerted people to the truth: teaching reform and innovation can be effective only under the guidance of the elicited thinking of teaching (Huiyu Dong, 2000).

In the process of the application of the method, two key links should be taken into account. One is the choice of elicited contents. The elicited contents should have the characteristics of reflecting the essential relationships and law of things, representing the structural system of the subject frame and its basic theory. The teaching should embody the typical nature of complicated phenomena and trends of things, and should present difficulties to, and stimulation of, the students' intellect. The other key link is the eliciting question to be put forward. The question is at the heart of the elicitation method. Firstly choose the appropriate subject for elicitation, taking into account the difficulty, depth, span and gradient of the eliciting question which should be representative, typical and a good example. Secondly choose the proper style of representing questions. Thirdly design carefully the thinking, process and structural style of the question to be researched and create a good situation to discuss, deduce and explore the question as far as possible. Finally briefly introduce and summarize the question, arrive at the true conclusion, sum up the thinking and method of research, and try to let students write about their experience and summary of the research.

Mathematical modelling includes the following repeated process: the objective fact is abstracted and simplified; related variables and parameters are clarified; according to "some law" the mathematical relationships between variables and parameters are established; the solution is given analytically or approximately, and the explanation and verification are added to the solution.

The mathematical modelling-orientated teaching method of elicitation is a teaching method of elicitation with the elicitation problems solved by mathematical modelling and the elicitation content of thinking, technique, knowledge and means of mathematical modelling. The processes are:

a. Creating the setting.

That is, students are led into the "setting" by means of creating the right atmosphere. The approaches used include forming a connecting link between what comes before and what goes after, stimulating interest by telling stories, summarizing, coming straight to the point and offering difficult questions. For example, in the classroom teaching of the application of normal differential equations, the processes are as follows:

(Teacher) Please recall what kind of normal differential equations we have learned and how we solved them?

(Show) The table of common solutions of normal differential equations.

(Teacher) I ask you to learn it well and grasp it since we have finished it. How can you meet this need? Please look at the opinions of two famous mathematicians.

(Show) "What does it mean to learn mathematics? Of course we hope to use it well and the best use is using. You are learning really only when you do your own thinking, as is well known to all" (by Whitney Hassler, a famous contemporary mathematician, educationalist and winner of the Wolf prize).

"Mathematics begins with reality and is used in reality" (by Hans Freudenthal, a famous mathematical educationalist).

b. Putting forward the problem.

According to the teaching content, students' majors and practical questions, the questions with proper difficulty are put forward and students' thinking is stimulated under teachers' induction. For example, in the classroom teaching of the application of normal differential equations, the processes are as follows:

(Teacher) Now let's look at a problem:

(Show)   PROBLEM: CAN   COASTGUARD   VESSEL   SEIZE SMUGGLER (Xinsan Li, 1997))?

In Fig.1, travelling at maximum speed v_0 along a line parallel to the OY axis from A on OX axis, the smuggler was spotted by a coastguard vessel at the origin O on the coast. The coastguard vessel pursues the smuggler rapidly, at speed $2v_0$. Where will the coastguard vessel seize the smuggler? If OA = 6miles, and v_0 = 14 miles per hour, when will the coastguard vessel seize the smuggler?

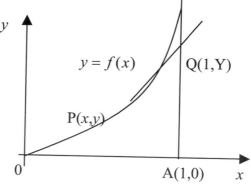

Fig.1

c. Analyzing the problem.

The known condition is found out; the unknown condition is made clear; the appropriate hypothesis is put forward and the solution strategy is chosen. For example, in classroom teaching of the application of normal differential equations, the processes of analysis are as follows:

(Writing on the blackboard)

a) Here we think the smuggler can be seized when the coastguard vessel and the smuggler are at the same place at the same time;

b) the key of the solution is to find the paths taken by the two vessels;

c) if OA=1, the smuggler is at Q(1,Y) when the coastguard vessel tracks P(x, y). The tracking curve of the coastguard vessel is $y = f(x)$. Because the tracking direction of the coastguard vessel is toward the smuggler at all times, PQ should be a tangent to the tracking curve. If the relationship between speeds is combined, $y = f(x)$ is established.

d) Constructing the model.

By the correct simplification and the use of "law", the translation of "question" is finished; the question is restated by using the "language" of mathematics, and the mathematical modelling of question is completed. By using mathematical knowledge, the mathematical model of above problem is

$$\sqrt{1+y'^2} = 2(1-x)y'' \text{ with } y(0)=0 \text{ and } y'(0)=0 \qquad (1)$$

e) Solving the model.

By using mathematical knowledge, the computer and proper software package, and by analyzing, calculating, graphing and testing, the mathematical model above is solved. The particular solution of the above differential equation is

$$y = -(1-x)^{\frac{1}{2}} + \frac{1}{3}(1-x)^{\frac{3}{2}} + \frac{2}{3} \qquad (2)$$

f) Explaining the solution.

The practical meaning of the mathematical solution is expounded and interpreted in terms of the real world problem. The tracing curve of coastguard vessel is given by equation (2).

Let $x = 1$, then $y = \frac{2}{3}$. At $\left(1, \frac{2}{3}\right)$, the smuggler should be seized.

As OA = 6 miles, and v_0=14 m/h, the time taken to seize the smuggler is

$$T = \frac{\frac{2}{3}OA}{v_0} = \frac{\frac{2}{3} \times 6}{14} \approx 0.286 \approx 17 \text{ minutes and 9 seconds}$$

g) Discussing and extending.

Students are encouraged to think about and discuss such questions as the appropriateness of the solution, the effect of the change of initial conditions on the solution and "similar" questions in real life. For example, for the problem above, the discussion could be as follows:

(Teacher) Until now we have solved all the problems, but we can think further.

(Show)   STUDIES:

a) Can coastguard vessel seize the smuggler if they have the same speed? Why?

b) Are there other plans for intercepting and capturing than the plan of the pursuit curve?

c) Are there similar problems in real life? Please raise several examples.

(Give the keys to the studies above after directing and discussing)

                    KEY to STUDIES:

a) Can not. Use *reduction ad absurdum*

$$T_c = \frac{D_c}{v_0} \geq \frac{\sqrt{1+Y^2}}{v_0} > \frac{Y}{v_0} = \frac{D_o}{v_0} = T_o$$

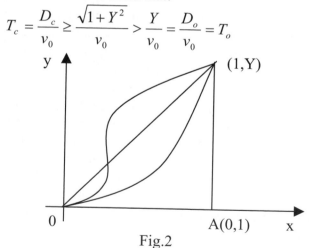

Fig.2

Here: T_c = the time that the coastguard vessel has taken

D_c = The distance that the coastguard vessel has covered

T_o = The time that the smuggler has taken

D_o = The distance that the smuggler has covered

b) Yes. The coastguard vessel can bear up to (1,Y), tracing the beeline. Then we can solve

$$\frac{\sqrt{1 + Y^2}}{2v_0} = \frac{Y}{v_0} \qquad \text{which yields: } Y = \frac{\sqrt{3}}{3}$$

c) Yes. For example: a missile chasing a target, a cat chasing a mouse (a cheetah preys on animals and a hungry tiger pounces on animals) etc.

There are some laws in the teaching but the laws are indeterminate, because everything is in a state of change. If a mathematical modelling way of thinking is embedded into the teaching, the principle of elicitation adhered to and students' enthusiasm aroused positively, then the process of their learning mathematical knowledge is changed into the process of them acquiring knowledge by the use of their own mind. Their ability to think, acquire new knowledge and solve problems on their own is improved. And this meets the requirements of the mathematical modelling elicitation method of teaching.

3. Some important issues on the mathematical modelling-orientated teaching method of elicitation

3.1 Topics applicable

Any method is used in certain situations. The best topics used by the mathematical modelling-oriented teaching method of elicitation are the Maximum or Minimum of Functions, Applications of Integrals, the Maximum or Minimum of Multi-variable Functions, Applications of Power Series, Applications of Fourier Series, Application of Differential Equations. Certainly, these methods can be transferred to other topics. For example, a good problem to stimulate students' interest can be given when a new definition or new content is introduced, and the applicability of the content is stated so as to arouse their thirst for knowledge. The question can be illustrated by a graph, and applications in studies closely related with their major subject should be increased, combined with the course of advanced mathematics. Five or six topics can be chosen to give students a practical assignment, and they are required to finish the process on their own by using their knowledge and skills to answer some simple questions.

3.2 Essence of the method

To adhere to the principle of elicitation all the time in teaching is the key to success of handling the mathematical modelling-oriented teaching method of elicitation. For this reason, teachers should pay more attention to the following in their teaching:

a.   Treating students wholeheartedly and honestly in order to bring their dependence, initiative and creativity into full play;

b.   Creating an air of comfort, harmony, learning from each other and the courage to explore questions so as to bring the study group into full play;

c.   Encouraging them to unite and help each other so that they could demonstrate their ability and develop themselves;

d.   Attaching great importance to students exploration, guiding them in comprehending questions as a whole, encouraging them to guess boldly, and widely using the way of analysis, comprehension, generalization, specialization, deduction, induction, analogy and association, laying equal stress on thinking in images and logical thinking, and changing the trend of emphasizing well-organized deduction to the neglect of intuitive guesses;

e.   Re-enforcing their belief system in solving problems, arousing their interest in solving problems by revealing mathematical characteristics of simplification, symmetry and unity, and cultivating their psychological qualities of attention, concentration, persistence, flexibility, motivation and attitude;

f.   Guiding and encouraging them to explore (model, characteristics of numerical value and algebra, and function and test mathematics) thinking by means of electronic calculator and computer, to improve their skill (evaluation, calculation, graphing and analyzing data), and get the more advanced mathematical thinking and experience which can not be finished by means of traditional pen and paper;

g.   Setting up an open classroom and attaching importance to students, positive study by changing the traditional teaching method (teachers explain content while students listen to teachers, do exercises that are corrected by teachers, and take examinations), letting them understanding the learning process and develop the way of thinking, while at the same time they acquire knowledge, and change "learning to understand" into "understanding to learn", thus gaining training in learning strategies;

h.   Paying attention to the diversification of the elicited style and making good use of the style as long as it can meet the needs of the elicitation method of teaching.

3.3 Teaching conditions

The basic conditions of handling the mathematical modelling-oriented teaching method of elicitation are selected mathematical modelling case studies and mathematical software. Nowadays society has entered the Internet information age, and modern educational technology in which computers play a key role has been used widely. The rapid development of numerical calculation, symbol calculation, dynamic sound and picture, and large-scale mathematical software packages has strongly extended the time and scope of teaching. This plays an important role that cannot be replaced in helping students understand deeply mathematical thinking and to complete calculations quickly, adding teaching information, improving teaching effect and quality, and advancing teaching reform.

Supplementary teaching software and deep interest are the necessary pre-requisites for using the based mathematical modelling elicitation method of teaching in classroom. Therefore, the development of computer-aided instruction software concerning the study of Advanced Mathematics and teaching demonstration must be greatly strengthened. At the same time, the development and the collection of teaching material regarding mathematical modelling must be emphasized and the old must be weeded through to bring forth the new. The studies in every mathematical modelling teaching module are the finishing touch, so every effort must be made to set the studies.

References

Huiyu Dong (2000) 'The Course of Pedagogy in Contemporary Military Universities' Beijing: Military Science Press, 262-268.

Qixiao Ye (2001) 'Guidance Materials of Mathematical Modelling Competition between University Students (IV)' Changsha: Hunan Educational Press, 8-13.

Ruiping Hu (1996) 'Mathematical Modelling and the Reform of Mathematical Education in Military Universities' Army Watercraft, Periodical 4

Xinsan Li (1997) '205 Applied Examples of Advanced Mathematics' Beijing: Higher Education Press, 221-223.

23

Teaching and Assessment of Mathematical Modelling in Community Colleges

Lu Xiuyuan, Mo Jingjing and Lu Keqiang
Feng Tai Workers and Staff University, Beijing, China
lyhao@crsc.com.cn

Abstract

Since the 20th century, Mathematics has been applied widely in all of the important branches of learning. Mathematical modelling, due to its wide participation, has greatly promoted a close integration of science, technology and economics. In our country, the activities of the mathematical modelling competition have developed rapidly in colleges and universities but the teaching of mathematical modelling is still progressing rather in community colleges. In 1996, on the basis of the research into the "three-dimension teaching mode" in adult education, Feng Tai Workers and Staff University offered a new course of mathematical modelling, which provided a successful exemplar for community colleges in the Beijing area.

1. Mathematical Modelling Teaching Experiment

Mathematical modelling is a mode of thinking and a process of doing in which a specially designated problem is analysed and a mathematical structure is built by using modern mathematical theory. [Since 1980, mathematical modelling has linked science, technology and economics due to its application in many fields. Universities have given courses in mathematical modelling for many years, and this has played an important role in improving student's qualities, in fostering their innovative and practical abilities]. In comparison, although students in adult colleges have the advantage of practical experience, they cannot be given a course of mathematical modelling due to the constraint of the conventional teaching mode. Over a long period of time, the problem of whether the course should be given and how it should be given in adult colleges (community colleges) has become a new research task for educators of mathematics both at home and abroad.

1.1 Problem Raised

Since 1988, in the course of doing research into the three-dimension teaching mode for adult education, we realized that there is an internal and essential relationship between mathematical modelling teaching and three-dimension teaching mode, for we attended several seminars on mathematical modelling organized by Beijing Mathematical Association University Committee. Mathematical modelling required a new field which linked teaching closely with practice. The three-dimension teaching mode experiment required "two-way participation of both teachers and students," which asked for close contact with enterprise's real situation. Students, by way of designing a reforming scheme for their own enterprises, helped their enterprises to make profit. For mathematical modelling, the three-dimension teaching mode is, in essence, a mathematical modelling system. The reforming schemes designed by students for their own enterprises are a series sub-systems of mathematical modelling. Each particular scheme is a practical valuable mathematical model. So it is not only to benefit the research, but also to benefit the enterprises that we introduce the mathematical modelling course during the research in combination with the three-dimension teaching mode for adult education.

1.2 Theory Demonstration

At the beginning of 1996, we gave a report on developing mathematical modelling teaching in Beijing adult colleges. And at the end of that year, the report was presented both in the 33rd Seminar of Mathematical Modelling organized by Beijing Mathematical Association University Committee and in Beijing Mathematical Association Annual Meeting held in Qing Hua University. Many experts and scholars gave their opinions about it.. The article titled *Thinking on Mathematical Modelling Teaching and Application in Adult Colleges and Universities* (published in *Practice and Knowledge of Mathematical 27th volume 1997*) was the theoretical condensation of the report. After the examination given by the experts of the Institute of Beijing Municipal Educational Science and the approval of the Adult Education Department of Beijing Municipal Education Commission, we offered, for the first time, the course of Mathematical Modelling to the students of "Rural Enterprise Class" (which refers to the subject of rural enterprise management) of Grade 96 (entered in 1996) and Grade 97 (entered in 1997) in Beijing Feng Tai Workers and Staff University and it proved successful. Two factors were taken into consideration in choosing the classes for the experiment. One is that the 'Rural Enterprise Class" was a new subject in Beijing Adult Colleges which appeared in 1995, so that the experiment might be used to examine the applicability of mathematical modelling teaching for the new subject. The other is that, in comparison with students in other specialties, the elementary knowledge of the students in this class is rather poor. The statistics obtained from Beijing General Mathematics Examination with a common test paper for all students from different schools and fields shows that less than 50% of the students in Grade 95 passed the exam. The experiment may examine the adult students at different levels on how much they participate in mathematical modelling teaching in an all-round way.

1.3 Experiment Process
1.3.1 Investigation on the Spot.

The superiority of community colleges is their "natural experiment places" which is a beneficial condition for converting scientific technology into productivity in the process of adult education. And in the process of the education reform experiment carried out in the "Rural Enterprise Classes", we took advantage of this superiority to make an overall investigation of the students' working units. 34 symposiums were given without adding any burden on their enterprises, 272 class-hour practice in spare time was accomplished. In the investigation, we collected, from all sides, ideas, opinions, data and materials on the present condition and problems. We came to understand their urgent need for mathematics applied to economics due to the development of their enterprises, and all of this laid a solid foundation for data analysing and processing, mathematical model building and correct decision making.

1.3.2 Making outline.

We offered the new course of *Mathematical Modelling* with such modern mathematical knowledge as *Input and Output, Orthogonal Experiment, Network Analysis, Decision Making Theory* etc. on the condition that the original class hours for *Basis of Economic Applied Mathematics* were not reduced. At the same time, teachers and students together built the mathematical models, designed reforming schemes and solved practical problems according to the situation of the students' enterprises. In this way, we made the mathematical modelling teaching suitable and practical for adult students on jobs and met the need of social economic development.

Teaching outline:

Chapter One------Quantitative Analysis and Mathematical Modelling

Chapter Two------Differential Equation and Input & Output

Chapter Three-----Property Assessment and Convert into Share

(Including special knowledge, experience, skills, etc.)

Chapter Four------Forecasting Methods and Decision Making Types

Chapter Five------Orthogonal Experiment and Network Analysis

Chapter Six--------Numerical Value Calculation and Financial Control

The key point of the outline was the integration of theory and practice.

The aim of the course was to improve the students' ability for analysing and solving practical problems by studying mathematical modelling.

1.3.3 Experiment Result.

There were 68 students in the "Rural Enterprise Class" of Grade 96 and Grade 97. Having attended the course, the students' thinking was inspired, their interest in taking part in scientific researches were aroused, their ability to doing scientific research were fostered and their knowledge was built up. They read 99 reference magazines, 74 relevant specialized books and set up 61 books of reading cards in their spare time. They took part in the Beijing General Mathematics Examination and passed the examination. Their average score is 88.9 and one of them scored 100. The case studies analysing their enterprises' actual situation written by them were published in *Case Analysis on Medium-small-sized Enterprise Reform and Development* (Beijing Normal University Publishing House Mar. 2000). In their papers, they constructed the mathematical models and put forward reforming schemes. 133 papers were written on case analysis in all and 57 of them were implemented. The implementing rate was 42.8%. By the end of 1999, the enterprises had made profit of 18,430,000 yuan. The achievement won the first-grade-prize of Adult Education Training Engineering Project in December 1999, which was awarded by Beijing Municipal Education Commission. It also won the second-grade-prize of Teaching Achievement in December 2000, which was awarded by Beijing Municipal Government and the first-grade-prize of Teaching Achievement in January 2001, which was awarded by Feng Tai District Government.

1.3.4 Some Cases of Mathematical Modelling

The characteristics of adult students made them not only the objects of the experiment but also the participants in the whole teaching process. The extent to which students participated in the experiment can directly influence the results of the teaching experiments. The case-analyses written by them involved 35 aspects. They are involved in up-grading rural enterprises, developing new products, modifying old products, reducing production costs, rationalising uses and development of land, inviting overseas investment and businesses, establishing joint-stock systems, optimising transportation management, predicting production, making decisions, financial analysis and controls, and old village reformation. The followings are some of the examples:

Case 1: *Control the Elements of Input and Output and Increase Social Economic Profit* written by Dai Pengfei (Grade 96).

In the paper he collected the data of the income and the expenditure of the Yue Ge Zhuang Farmers' Market in Feng Tai District from 1991 to 1997. He set up an input and output mathematical model and put forward a scientific management program which had been put into effect. As a result, the income of 1998 increased by 25%, which was 2 million yuan more than the previous year.

Year	Number of staff	Annual income (¥10thousand)	Annual expenditure (¥10thousand)	Expenditure as % of income	Profit	Profit as % of income	Income ratio increase
91	112	294	134	46%	160	54%	——
92	160	385	195	51%	190	49%	31%
93	286	434	239	55%	195	45%	13%
94	286	725	556	77%	169	23%	67%
95	286	1112	531	47.7%	581	52%	53%
96	283	1256	507	40%	749	59%	13%
97	283	1250	850	68%	400	32%	-0.5%

Table 1. Economic Data Analysis of the Farmers' Market (1991-1997)

expenditure income		expenditure in					Total expenditure
		fruit dep	vegetable dep	other dep	total	other items	
Source of income	fruit dep	X11	X12	X13	N1	S1	X1
	vegetable dep	X21	X22	X23	N2	S2	X2
	other dep	X31	X32	X33	N3	S3	X3
add up to		X1	X2	X3			
profit after tax		M1	M2	M3			
total income		Y1	Y2	Y3			

Table 2. Balance Table on Income and Expenditure

(The concrete input and output data of the table has been omitted accordingly)

The Model Analysis: The Balance Table on Income and Expenditure can be set up according to an input and output mathematical model. The direct consumption elements matrix of the market can be obtained. From these, each department will not only know the ratio of income and expenditure of itself but also know that of other departments. For example, suppose the direct expenditure of the vegetable department in 1991-1997 is a_{11} = 0.46, 0.51, 0.55, 0.77, 0.47.7, 0.40, 0.68 respectively. According to the direct consumption element matrix, the manager can further analyze the detailed financial account to find out ways to reduce costs and make decisions. Meanwhile, the manager can also forecast next year's income and expenditure according to the matrix in order to obtain the scientific management results of the combination of qualitative analysis and quantitative analysis.

If we forecast the income of 1997 according to average growth rate, it should be:

$$X_{t+1}=X_t+X_t (t+t_{-1}+t_{-2}+----------+t_{-n+1})/n$$
$$=1256+1256 (0.31+0.13+0.67+0.53+0.13)/5$$
$$=1256+1256 (0.354)$$
$$=1901 (¥10 \text{ thousand})$$

Suppose the annual growth rate is 0.25, then the income of 1997 should be:

$$X_{t+1} =X_t+X_t (0.25) =1256+1256 (0.25) =1570 (¥10 \text{ thousand})$$

From above, we can forecast that the income of fruit and vegetables in 1997 should be 1570---1901 (¥10 thousand), which should be 320---651 (¥10 thousand) more than the actual income.

2. Evaluation of the Mathematical Modelling Teaching Scheme

The teaching method of mathematical modelling in community colleges adopts the three-dimension teaching mode. The ability to solve practical problems is used as the assessing standard and form. In our teaching experiment, the students are required to find out problems from their practice, build mathematical models accordingly, put forward and theoretically prove the reform scheme. The examination form is to accomplish an economic applied paper. The assessing standard is to see how well the students can put their reform scheme into production practice and how much their scheme can be used in the production practice.

Three dimension teaching mode for adult education (the abbreviation for which is three dimension teaching mode) is the achievement in scientific research of the adult educational teaching reform of Beijing. To put it briefly, it is a kind of teaching mode of organically putting "theoretical study", "production practice" and "education result" together in the whole process of adult education. Three dimension teaching mode comes from practice and is tested by practice. It was put forward and designed on the basis of economic applied mathematics teaching reform, which was carried out in Beijing TV University Factory Director and Manager class from 1988 to 1989. This was once more proved by the teaching reform carried out in an accountancy class and an enterprise Management Class of Feng Tai Workers and Staff University from 1990 to 1993. On the basis of the above work the research group designed the structural sketch of the mode. The Appraisal Committee of Beijing Adult Education Bureau attached great importance to it. On the basis of 12 years of teaching reform experiment, the achievement of case teaching in three dimension teaching mode titled "Case Analysis on Reform and Development of Medium-small-sized Enterprises", was published in March 2000. It widely aroused social influence and won the second-grade prize of Teaching Achievement awarded by Beijing Municipal Government.

Three dimension teaching mode is a pioneering achievement in academic research of adult education. It puts forward and theoretically proves the following original and important academic problems.

(1) It has built systematic mathematical models which take social economic development and the need of each student's individual development as its original point, and set up the third dimension coordinate axis Z (education results) on the basis of two-dimension axes of X (theory learning) and (production practice). All of which forms a three-dimension teaching mode with its own solid structure. It takes "education result as the premise and the test standard for the development of adult education.

(2)The new idea of the theory is that it examines the "education result" from a completely new point of view. It tries to improve the education result throughout the whole education process, it solves the difficult problem of uniting theory with practice in adult education and theoretically proves that education result can be produced in the process of adult education.

(3) The practical new idea of the theory is that it takes advantage of "natural experiment places" and brings students and their working units into the whole education system. Both the teachers and students take part in the experiment and under the guidance of the teachers the students put what they have learnt into production practice and in this way the production result is obtained. The design structure of the three dimension teaching mode has laid the foundation for uniting theory with practice.

Quantitative Index of axes X, Y and Z

On the basis of this understanding and through further research and experiment, we have accomplished the design for the axes of X, Y and Z, which is as follows:

(a>0, b>0)

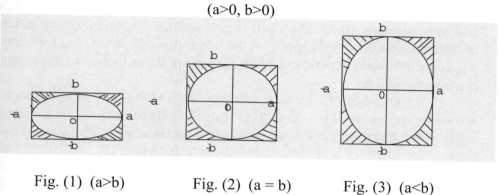

Fig. (1) (a>b)   Fig. (2) (a = b)   Fig. (3) (a<b)

The quantitative standard of axes X and Y is designed according to the proportion of the rectangle (square) and the ellipse (circle). Suppose: the area of the rectangle (square) is $S = 2a \times 2b = 4ab$. The area of ellipse (circle) is $S_1 = \pi \times ab$, the area of the shaded part is $S_2 = S - S_1 = 4ab - \pi ab = (4-\pi)ab = 0.86ab$

The purpose of this design is to use the least input to produce the biggest education result under the condition of not increasing the original class hours.

Suppose: u ($u \geq 1, u \in N$) is the original class hours, N is a natural number.

The coordinate of axis X is αu,

The coordinate of axis Y is βu,

$$y = \frac{\text{the numer of schemes already put into effect}}{\text{the number of schemes put forward}} = \frac{57}{133} = 0.428$$

The coordinate of axis Z is γu

$$\alpha = \frac{S_1}{S} = \frac{\pi ab}{4ab} = 0.785, \quad \beta = \frac{S_2}{S} = \frac{0.86ab}{4ab} = 0.215$$

The quantitative index of three-dimension teaching mode of axes X, Y and Z is

$$X = \alpha u = 0.785u, \quad Y = \beta u = 0.215u, \quad Z = \gamma u = 0.428u$$

For example: When the original planned class hours u = 80 hours;

The class hours for theory learning X = 63 hours

The class hours for production practice y = 17 hours;

The education result class hours produced Z = 34 hours;

From the above we can see that 63+17+34 > 80, which reflects the principle that educational management produces profit, i.e. 1+1>2.

Note: γ = 0.42 refers to the coefficient of teaching reform and experiment during the ninth five-year-plan. The value during the eighth five-year-plan is γ =0.35. The coefficient γ on axis Z is an important index of education result. The bigger γ is, the bigger the education result will be. γ is concerned with experiment items. The more items there are, the bigger the figure γ will be.

The statistics of education result belongs to the field of educational economic research. The education result of three-dimension teaching mode is produced on the basis of not increasing education funds and class hours. So there is not any contradiction between the statistics and measurement of education result and the other methods of the existing educational economics. The new idea of it is to combine theory learning with production practice through the operating system of teaching mode, and to obtain optimum results and continuous development under the condition of the original manpower and the original material capital.

2.1 Analytic Diagram of Education Efficiency

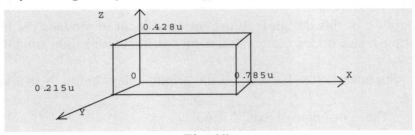

Fig. (4)

Practical education result $\omega_i = \alpha_i\beta_i\gamma_i u$

Analytic formula of education efficiency $\delta = \dfrac{\omega_i}{\omega_0}$

Input education cost $\omega_0 = 0.785 \times 0.215 \times 0.215 \times u^3$

Practical education result $\omega_1 = 0.785 \times 0.215 \times 0.428 \times u^3$

During the ninth five-year-plan period:

Education results index, refer to Fig (4)

Practical elastic index= $\gamma-\beta$=0.213

Practical index $\beta_1=\beta+(\gamma-\beta) = \gamma = 0.428$ (a>b)

As the same theory: Potential elastic index =$\alpha-\beta$=0.57

Potential index $= \beta_2 = \beta + (\alpha-\beta) = A = 785$ (a=b)
Maximum elastic index $= \alpha-2\beta+\gamma = 0.213+0.57 = 0.783$
Maximum potential index $\beta_3 = \alpha-\beta+\gamma = 0.215+0.783 = 0.998$ (a<b)

☐ Practical education efficiency $\delta_1 = \dfrac{\omega_1}{\omega_0} \approx 1.99$

Practical education efficiency $\delta_2 = \dfrac{\omega_2}{\omega_0} \approx 3.96$

Potential education efficiency $\delta_3 = \dfrac{\omega_3}{\omega_0} \approx 7.22$

Maximum potential efficiency $\delta_4 = \dfrac{\omega_4}{\omega_0} \approx 9.24$

From the above diagram and formulas, we can see that when the index $\beta = 0$, the index of the theory in combination with production practice is 0. This mathematical formula can be used to check the combination of theory and practice. If $\beta = 0$, then γ must be 0, $\omega = 0$.

It shows that the quantitative design of three-dimension teaching mode is rational, which gets rid of the shortcoming of only having the integration of qualitative analysis but without having the integration of quantitative examination.

The education efficiency of three-dimension teaching mode is bigger than 1, i.e. $\beta > 1$, which shows that the design of three-dimension teaching mode is scientific. According to the comparative research over 12 years on the three-dimension teaching mode carried out between a teaching reform experimental class and a non-teaching reform experimental class, we have found that the students in a non-teaching reform experimental class also have the initiative to solve problems by applying theories to reality on their own because of their feature of being in jobs. Even some adults through self-study can also have this kind of initiative. But the education efficiency produced by means of this method is less, i.e. $\beta < 1$. The index γ on axis Y is less than 0.215, i.e. $\gamma < 0.215$. On general condition, $\gamma < 0.10$, the produced results of which are largely at random. Since three-dimension teaching mode is guaranteed by its operational system, and for all the students so the education result is inevitable. That is why the education efficiency $\beta > 1$, Y-axis index $\gamma \geq 0.215$. The value of β is the checking standard for the implementation of three-dimension teaching mode. It is also the assessment index for education result.

2.3 Education Results and Marginal Education results Function

Suppose: Education result function $W_A = A \, LnV$,

Marginal education result $\dfrac{dW_A}{dV} = \dfrac{A}{V}$,

Among which: A refers to practical economic efficacy and V refers to the number of students who attended in the course by using three-dimension teaching mode. The so-called "marginal education result" means that when the other education elements (leading cadres, experts, teachers) are not changed, the education result increased by adding one unit (here refers to one student) to student quantity V.

According to the same theory:

Suppose: B is potential economic profit,

Potential education result function $W_B = B \, LnV$,

Potential marginal effect $\dfrac{dW_B}{dV} = \dfrac{B}{V}$

Both the direct and potential economic profit in the eighth five-year-plan are more than ¥7 million and that of which in the ninth five-year-plan are more than ¥18 million and ¥ 28 million respectively.

In "the eighth five-year-plan" period:

$\dfrac{dW_A}{dV} = \dfrac{A}{V} \approx 10.4$ (¥10 thousand): $\dfrac{dW_B}{dV} = \dfrac{B}{V} \approx 10.4$ (¥10 thousand)

In "the ninth five-year-plan" period:

$\dfrac{dW_A}{dV} = \dfrac{A}{V} \approx 26.5$(¥10 thousand): $\dfrac{dW_B}{dV} = \dfrac{B}{V} \approx 41$(¥10 thousand)

The extension of education result function $W = W_1 + W_2 + W_3 + W_4$

($0 < a_i < 0.25$, $\sum\limits_{i=1}^{4} a_i = 1$)Among which:

$W_1 = a_1 A \, LnV$, Social effect, $W_2 = a_2 A \, LnV$, Economic profit,
$W_3 = a_3 A \, LnV$, Personnel quality, $W_4 = a_4 A \, LnV$, Social personnel information network.

When three-dimension teaching mode is in normal progress, $a_i = 0.25$, each index should be balanced.

The connotation of education result function is usually recorded as $W = A \, LnV$, for LnV is a monotonic increasing function. So LnV will be increased progressively as the student's quantity V increases. In return, the education results will be bigger and bigger and all of the above shows the popularizing value of the three-dimension teaching mode.

2.4 Orthogonal Experiment Design of Education Element

It is obvious that the results of teaching reform experiment in the period of the ninth five-year-plan is better than that in the period of the eighth five-year-plan which has a close relationship with education environment. The mathematical model of education management in three-dimension teaching mode is a kind of orthogonal experiment consists of 4 elements at 3 levels. The 4 elements include leading cadres, experts, teachers and students. The optimum combination among them is the necessary condition for achieving the best education result.

Because of the complicated relationship among the four elements in reality, we set each element at 3 levels in our experiment on the basis of further investigation. The orthogonal experiment design has been made for the education result, which supplies scientific support for foreseeing, controlling, combining and selecting the optimum scheme of education result. The figures in table 4 are tallied with the result of teaching reform obtained in the eighth five-year-plan and once again verified in the ninth five-year-plan.

Element Level	A Leader at each level	B Related experts	C Teachers attending the experiment	D students attending the experiment
(1)	A_1	B_1	C_1	D_1
(2)	A_2	B_2	C_2	D_2
(3)	A_3	B_3	C_3	D_3

Table 3 Table of Orthogonal Experiment for Education Result

A refers to leaders at different levels (government, education department and students' working units) and their attitude towards teaching reform experiment. The 3 levels are (1) support positively A_1; (2) support A_2; (3) not support A_3. The actually existed "suppression and retaliation" is not included in the table. B and C respectively refer to the professional skills and attitude of the experts and teachers towards their work.

The 3 levels are that (1) B_1 and C_1 are the best; (2) B_2 and C_2 are better; (3) B_3 and C_3 are general. (The other combinations are not included in the table, in implementation conservative and average values can be put into appropriate levels. The lowest level for B and C is just qualified.) D refers to that if the subject the students have learnt suits to their jobs and if they have decision-making rights in their working units. The 3 levels are (1). The subject they have learnt suits to their jobs and they have decision-making rights D_1; (2). The subject they have learnt suits to their jobs but no decision-making rights D_2; (3). The subject they've learnt does not suit to their jobs D_3; (decision-making right is not designed for D_3).

Element Serial no	leader at levels A	Related experts B	Teachers attended C	Students attended D	Result obtaining ratio %
1	(1)	(1)	(1)	(1)	100
2	(1)	(2)	(2)	(2)	83.75
3	(1)	(3)	(3)	(3)	67.5
4	(2)	(1)	(2)	(3)	70
5	(2)	(2)	(3)	(1)	72.5
6	(2)	(3)	(1)	(2)	71.25
7	(3)	(1)	(3)	(2)	58.75
8	(3)	(2)	(1)	(3)	57.5
9	(3)	(3)	(2)	(1)	60
$\sum X_{1J}$	251.25	228.75	228.75	232.5	
$\sum X_{2J}$	213.75	213.75	213.75	213.75	
$\sum X_{3J}$	176.25	198.75	198.75	195	
$\overline{X_{1J}}$	83.75	76.25	76.25	77.5	
$\overline{X_{2J}}$	71.25	71.25	71.25	71.25	
$\overline{X_{3J}}$	58.75	66.25	66.25	65	
R_J	25	10	10	12.5	

Table 4 Table of orthogonal Experiment for Education Results

Table of orthogonal experiment for education result is designed by means of $L_9(3^4)$ as shown in the table. $\sum X_{ij}$ shows the sum of the experiment result of column J and level i. For example: column $A \sum X_{11} = 100 + 83.75 + 67.5 = 251.25$. $\overline{X_{ij}}$ shows the average value of the experiment index of column J and level I, such as

$$\overline{X_{11}} = \frac{251.25}{3} = 83.75.33 = 83.75.$$

Rj is called the difference which means the difference between the biggest of the three i.e. $\overline{X}_{1J}$, $\overline{X}_{2J}$, $\overline{X}_{3J}$ and the smallest of the three in the same column. For example, $R_1=83.75-58.75=25$. The index of difference Rj shows the influences in obtaining the mean index from the experiment. The bigger the differences are, the bigger the influence will be to the experiment index if the element is changed. By analyzing the Table of Orthogonal Experiment for Education Result we can draw the following conclusions:

Element A has the biggest influence on result ratio. When A is at level (1) and (2), the result ratio is over 70%; when A is at level (3), the result ratio is between 57.59% and 60%. For this reason the establishment of the environment for the education reform is a vital condition for success and leaders are the key points in establishing the environment.

Both elements B and C are as important as A in obtaining the result. If we study the level below the qualified, elements B and C will have more negative influence on obtaining the result. Consequently, to improve teacher's quality is a necessary for the implementation of three-dimension teaching mode.

Element D has a bigger influence on the experiment, which shows that the students attending the experiment are not simply to be experimented. In adult colleges, the implementation of three-dimension teaching mode should be fully connected to the concrete situation of each individual. The characteristics should be reflected in teaching students according to their own aptitudes and the participation of both teachers and students in teaching reform and in the experiment is an important change in teaching mode.

2.5 Analysis to Structure Diagram on Three-dimension Teaching Mode

The operation system of three-dimension teaching mode refers to the operating procession and action mode between the teaching and practicing of three-dimension teaching mode system in relatively corresponding stage. Depended on the five stages which developed gradually further in teaching and practicing and the participation by both teachers and students, a new idea of three-dimension teaching mode is presented and a thought of running modern adult high-learning schools with its own direction and training target has been reflected totally.

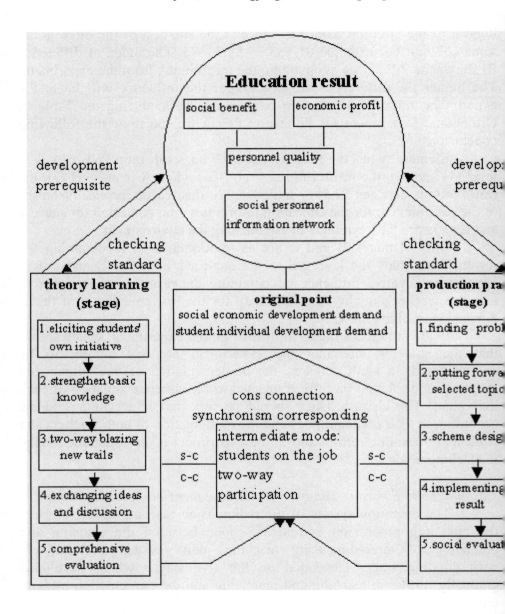

Offering mathematical modelling course in community colleges suits the characteristic of adult students who are on the jobs. It can make them to apply the knowledge to their production practice, extremely arouse their interests in combining what they have learnt with what they have done in their work. And in this way, it arouses the students' potential in bringing forth some new ideas so that the optimum results of education can be obtained.

Bibliography

Branch Group of three-dimension Teaching Mode for Adult Education (2000) Capital Normal University Publishing House.

Jiang Qiyuan (1996) 'Mathematical Models' Higher Learning Education Publishing House, 1-25.

Lu Xiu Yuan (1994) 'Theoretical Discussion on Three Dimension Teaching Mode' *Beijing Adult Education*, 11, 15-16.

Lu Xiu Yuan (1994) 'Practice and Investigation on Three Dimension Teaching Mode' *Beijing Adult Education*, 12, 12-13

Three-dimension Teaching Mode Case Teaching and Researching Group for Adult Education, 'Case Analysis on Reform and Development of Medium-small-sized Enterprises'.

Ye Qixiao (1998) 'Guidance Material for Competition of College Students Mathematical Modelling (Book 1)' Hunan Education Publishing House, 1-20.

24

The Role of Mathematical Experiment in Mathematics Teaching

Jinyuan-Li
Mathematics Group, Beijing No.4 High School, China
lijy22@yahoo.com, lijy77@sina.com

Abstract
    Nowadays the role of computers extends the means of solving problems. It inspires researchers' scientific insight and thus induces further challenges. After the Four-Colour Theorem was solved, many people began to conduct experiments on computers in order to support their mathematical theory. In developing and learning mathematics, the inductive and experimental method has raised the interest of academia again. In teaching mathematics, we not only need to train students' ability in logical reasoning, spatial sense and calculation, but also we need to foster students' ability in mathematical modelling, exploring and analyzing data. In other words, we need to improve their ability in "applying mathematics". Moreover, "mathematical experiment" is just an eligible teaching approach to serve such a purpose.

    In this paper, we will specialize in the importance of mathematical experiment and carry out an experiment which aims to identify powerful ways of implementing problem solving in the classroom using modelling and experiment as the focus of the activities proposed to students. The paper will attempt to elicit

       1.   The necessity of mathematical experiment;
       2.   The availability of mathematical experiment;
       3. A practicable teaching approach on mathematical experiment;

4.   *Students' discoveries;*
5.   *The value and effectiveness of mathematical experiment.*

1. Introduction

In this global information age, where both economic growth and individual opportunity are based on ideas, a commitment to giving all students the opportunity to develop their full potential and originality is vital to their strength and success. We need to solve from the beginning problems in economics, society, environment, etc., and this raises a serious challenge to education. Education is now charged with preparing students for careers dominated by powerful information technology. Therefore, We should improve students' ability in applying mathematics.

In the following discussion, we will specialize in the importance of mathematical experiment and carry out an experiment which aims to identify powerful ways of implementing problem solving in the classroom using modelling and experiment as the focus of the activities proposed to students. We will answer the following questions: How should teachers facilitate the development of students' creativity? Can we use the mathematical modelling method to solve mathematics problem? How can we use the computer in dealing with computationally approachable but otherwise intractable problems in mathematics?

2. The necessity of mathematical experiment

Mathematics nowadays is not only a type of science, but also an extensively applied key technology. As the information age is coming, the development of computer technology has a great influence on our daily life. The obvious function of a computer is to do enormous calculations at high speed, which makes it possible for those complicated, large scale problems. Highly computational, even "brute-force" methods were of necessity limited but the computer has changed all that. When the new generation of computers came out, the ways people learn mathematics and apply mathematics changed greatly. A re-concretisation is now underway. The mathematics problems can be transformed into a form suitable for a computer, even to the extent of verifying and understanding, thus new results or discoveries will be found. The computer-assisted proofs of the Four-Colour Theorem are a prime example of computer-dependent methodology and have been highly controversial despite the fact that such proofs are much more likely to be error free than, say, even the revised proof of Fermat's Last Theorem. (Jianwu-Wang, 1996)

Mathematics is traditionally considered to be a rigorous regime, which proceeds along the following thread: **definition, conjecture, theorem, proof, corollary.** Therefore, in order to support their mathematics theory, many people begin to conduct experiments on computer. In developing and learning mathematics, the inductive and experimental method has raised the interest of academia again (Mount Holyoke College, 1997). Those who identify themselves as "experimental mathematicians" are founding a new way to research mathematics, that is to say, to make new discoveries by computer experiment. From those mathematicians' points of view, mathematics is being changed into an experimental science. Experiment has always been, and increasingly is, an important method of mathematical discovery. (Gauss declared that his way of arriving at mathematical truths was "through systematic experimentation".) Yet this tends to be concealed by the tradition of presenting only elegant, well-rounded and rigorous results. (Yusun-Xie and Zhirang-Zhang, 1999; Epstein et al, 1992)

On the other hand, from an educational point of view, the emphasis of mathematics education should be changed according to the development of computing applied to mathematics. We should put more emphasis on mathematical modelling and mathematical experiment. The main subject of mathematical modelling is more than likely based on a substantial background while the mathematical experiment meets a more abstract *mathematics* problem. Even though, mathematical experiment can be treated as a special kind of modelling because its aim is to explore the model or theorem in mathematics. In this way, some of the abilities, which we used to be proud of, such as powerful paper-and-pencil computation skills, would not be that valuable. What people mainly focus on is to set up an algorithm to solve problems, then ask computer to do them. All you need to do is to manipulate the computer. Thus, people will not need to deal with details; instead, they can focus their minds on more valuable and original activities.

So in teaching mathematics, we not only need to train students' ability in logic reasoning, spatial sense and calculation, but also foster students' ability in mathematical modelling and ability in data processing. In other words, we need to improve their ability in "how to use mathematics", and mathematical experiment is a very good teaching form to serve such a purpose.

3. The availability of mathematical experiment

Today, in dealing with mathematics problems, some symbolic algebra software has been developed, such as *Mathematica*, *Maple*, *MATLAB*, and *MathCAD*. It is useful for mathematics teachers to introduce those systems to students, which will arouse their curiosity and enthusiasm to explore the unknown and encourage them to be creative. Moreover, students will benefit a lot in their future lives and careers.

For example, in the course of using *Mathematica*, users can draw graphs or carry out a calculation according to their needs. The transferring agility makes some sophisticated problems so easy. And it is very easy to learn *Mathematica*, you will get to know the basic function about it (including numerical approach, symbol calculation and visualization by computer) after three to five hours' computer learning. There is no doubt that it will benefit high school students a lot in their mathematical experiment.

4. The execution of mathematical experiment

We now turn to a more concrete example of a mathematical experiment. Our aim is to find evidence pointing in one or the other direction. We are hoping to gain insight into the nature of the problem from an experimental perspective.

4.1 The guiding principle

Curiosity and originality are daily occurrences for the students, but somehow most of us lose the freedom and flexibility when we grow older. The need to follow "directions" and "do-it-right", plus the many societal constraints we put on ourselves, prevents us from using our creative potential. For this reason, to teach high school mathematics, including doing mathematical experiment by computer, we should make students more initiative, leave them more space to think, lead them to ponder problems critically, deal with problems creatively, and develop their scientific consciousness in research, rather than restrict them in a straight jacket. Most of the creativities of every specialty are stimulated from intelligent prompting, which comes out from the process that the conclusion concurs or not with his own experiences after he puts his professional knowledge in consideration (Ethridge, 1995).

Therefore, the guiding principles of mathematical experiment are grounded in the belief that teachers should orchestrate the classroom environment in ways that promote equity and social justice, and that teachers should model democratic education. We should lead students to consider problems in an initiative, active and critical way, solve problems creatively, and develop their scientific consciousness and abilities.

4.2 An executive way

On May 25, 2000, we practiced a research class about the property of the function $y = a\sin x + b\cos x$ $(ab \neq 0)$ in class B5, grade 1 of Beijing No.4 High School. The course is designed as follows:

We divided students into several groups. Their tasks are as follows:

1) Give any value for a, b as you like, and draw the functions' graphs in the computer, observe carefully about the character of those graphs, and note down the results.

2) Analysis the influence of those values on the graphs, and answer the following questions:

✓ Is the graph appearing on the screen the function we are familiar with? What is its expression?

✓ What parameters can you predict from the expression? Put them in Table 1.

✓ Ensure parameters' values after the given a, b values. Put them in Table 1.

3) What efforts should we take to verify our findings? (Hint: you can consider it from experimental and theoretical views.) You should describe your experiment result clearly, and make efforts to explore the regularities from data related to the problems.

4) Based on the experimental phenomena, you can raise your conjecture and think about the advantages by doing so.

5) Give your support based on experimental phenomenon and analysis the possible mathematics proof.

6) Write a report of the experiment.

Parameters	(a, b)values			

Table 1

4.3 Comments

1) The experiment fully indicates the mathematics educational theory by experiment and induction:

- set out from some practical examples (including examples designed by students themselves)
- do experiments in the computer
- find out the rules among them
- make abstractions
- verify and prove your abstractions.

2) All the questions raised here are open to give students space to consider upon, and make their imagination and creativity fly. In this case, we have ample experimental evidence for the truth of our identity and we may want to take it as something more than just a working assumption.

3) Writing is of key importance to foster students' creative minds. "Good writing is equal to good thinking", though to write well does not mean to think logically, but it could be a good proof of good thinking. Learning and improving writing skills is efficient and necessary. In a course of writing, you should try to clarify your thoughts and organise them logically. The standard for how well you express yourself is based on other people's understanding. It is to our loss that most of the mathematical community are almost always unaware of how new results have been discovered. (Epstein et al, 1992) We know that the results are true because we have gone through the crucible of the mathematical process and what remains is the essence of truth. But the struggle, adventure, insight and thought processes that led to the result are hidden. For now we have used restraint in the presentation of our results in what we hope is an intuitive, friendly and convincing manner.

5. Students' discoveries

After learning from joyful exploration of mathematics, the students wrote brilliant experiment reports and proposed many problems. Students enjoyed researching and exploring; their creative ability was brought into full play.

In their reports and presentations, the students all found the model that the function $y = a\sin x + b\cos x \ (ab \neq 0)$ could be transformed into $y = A\sin(\omega x + \varphi)$ (see appendix); these two functions have the same properties. Furthermore, some students made many serendipitous discoveries, for instance:

1) When the range of x take a bigger value, the graph of $y = \dfrac{1}{x}$ is almost

coincident with that of $y = \sin\dfrac{1}{x}$, why? Because $\lim\limits_{x\to\infty}\dfrac{\sin\dfrac{1}{x}}{\dfrac{1}{x}} = \lim\limits_{x\to 0}\dfrac{\sin x}{x} = 1$.

And   they   proved   this   conclusion   by   using

$\sin x < x < \tan x$ $(0 < x < \dfrac{\pi}{2})$ and the Squeezing Theorem.

2)   After   exploring   the   properties   of   the   function $y = a\sin x + b\cos x$ $(ab \neq 0)$, some students proposed that they could extend   this   function   to   a   more   general   form: $f(x) = a\sin(\omega_1 x + \varphi_1) + b\cos(\omega_2 x + \varphi_2)$, and analyzed its properties. For example, they found an efficient method to determine the period of this function in some restrained conditions. Moreover, they found the function $y = \sin x + \sin(\pi x)$ is not a periodic function and gave a rigorous proof.

3) They extended the form of the function $y = A\sin(\omega x + \varphi)$ to the following   forms: $y = A(x)\sin(\omega x + \varphi)$, $y = A\sin[\omega(x)x + \varphi]$, $y = A\sin[\omega x + \varphi(x)]$, and $y = A\sin(\omega x + \varphi) + T(x)$. They found: $A(x)$ effects the amplitude of the sine wave, $\omega(x)$ effects the frequency, $\varphi(x)$ effects the phase, and $T(x)$ determines the tendency of the wave. These conclusions can help to explain the three radio wave modulation methods: Amplitude Modulation, Frequency Modulation, and Phase Modulation.

4) They found and explained the phenomenon of "beats" in physics if we play the sound of such function $y = \sin(\omega_1 x) + \sin(\omega_2 x)$ (ω_1 is sufficiently close to ω_2) as $y = \sin(1000x) + \sin(1010x)$.

This type of serendipitous discovery must go on all the time, but it needs the flash of insight that will place it in a broader context. It is like a gold nugget waiting to be refined — without a context it would remain a curiosity. During experiment, students encountered or proposed some problems that they could not resolve by their limited knowledge, but at least they achieve some knowledge for advanced study.

6. Effects of mathematical experiment

From the description above, mathematical experiment is an eligible teaching approach to stimulate students to learn mathematics actively.

Active learning, in which students write essays or perform experiments and then have their work evaluated by an instructor, is far more beneficial than passive learning. Through mathematical experiment, they became the "master" (not "follower") of the computer; thus they could focus their energy on practical problems. As students reflect upon the mathematics, the algorithms applied, and the sense-making approach, mathematical insight is deepened.

From the teachers' point of view, mathematical experiment is a beneficial complement to the traditional teaching model. Learning mathematics needs a continuous exploring process, and teachers should provide such circumstances. Moreover, the teachers need some special training when they learn mathematics teaching. Thus during the teaching process, the teacher can give constructive suggestions when students explore the problems themselves. In the long run, it is very important that students should not remember only some mathematics tricks, but increase their ability of exploring and resolving problems and set up their innovative consciousness. Therefore, we should be conscious that teachers' teaching thoughts should be renewed and transformed from authority model of "lecture" into consultant model of "inspire". Thus students know how they should do and what they should do, the teacher could be really freed from the dull cycle of "lecture, exercise, review". In the short-run, it would take much time to lead students to study mathematics by experiment. Looking from long-term, students who have got this kind of training would not only study much more actively and conscientiously, but also they would have strong team spirit. It is difficult to achieve such effects by the traditional training model.

From the sponsors' point of view, by teaching of experiment mode, students' creativity and cooperative spirit were developed greatly. It is unnecessary to interfere with students; they can do well! Then the teacher is freed off and does teaching research and improves teaching level. Therefore, as a sponsor, we should be conscious that mathematics needs a computer laboratory, and we should create such an environment in which each student in his own way follows his curiosity where it leads him, develops his abilities and talents, pursues his interests and preferences. In this way, his potentiality will be fully developed and he will make a maximum contribution to society when he grows up.

In summary, mathematical experiment is a very important method to develop students' creativity. It is a whole new teaching model with a brilliant future. Mathematical experiment is a very important complement for mathematical modelling; it is not a route to certain knowledge. Rather, mathematical experiment is exploring!

References

Epstein D, Levy S and Llave de la R (1992) 'About This Journal' *Experimental Mathematics* 1, No. 1.

Ethridge D (1995) 'Research Methodology in Applied Economics' Ames, IA: Iowa State University Press

Jianwu-Wang (1996) 'Introduction to Mathematical Thinking Style' Hefei: Anhui Education Press

Mount Holyoke College (1997) 'Laboratories in Mathematical Experimentation' New York: Springer-Verlag Ltd.

Yusun-Xie and Zhirang-Zhang (1999) 'Mathematical Experiment' Beijing: Science Press

Appendix

Solution 1: According to the experiment findings, we guess

$$a\sin x + b\cos x = A\sin(\omega x + \varphi)$$
$$= A\sin(\omega x)\cos\varphi + A\cos(\omega x)\sin\varphi$$

If we let $\omega = 1$, $a = A\cos\varphi$, $b = A\sin\varphi$, then

$$A = \sqrt{a^2 + b^2}, \quad \cos\varphi = \frac{a}{\sqrt{a^2 + b^2}}, \quad \sin\varphi = \frac{b}{\sqrt{a^2 + b^2}}$$

So, if we obtained the appropriate coefficients (A, ω and φ), then the function $y = a\sin x + b\cos x \ (ab \neq 0)$ could be transformed into $y = A\sin(\omega x + \varphi)$.

Solution 2:

$$a\sin x + b\cos x = \sqrt{a^2 + b^2}\left(\frac{a}{\sqrt{a^2 + b^2}}\sin x + \frac{a}{\sqrt{a^2 + b^2}}\cos x\right) \tag{1}$$

Let $A = \sqrt{a^2 + b^2}$, $\omega = 1$, $\cos\varphi = \frac{a}{\sqrt{a^2 + b^2}}$, $\sin\varphi = \frac{b}{\sqrt{a^2 + b^2}}$, then

$$(1) = A\sin(\omega x)\cos\varphi + A\cos(\omega x)\sin\varphi$$

We obtain $a\sin x + b\cos x = A\sin(\omega x + \varphi)$.

25

Theory and Practice in Teaching Mathematical Modelling at High School Level

Qiu Jinjia
The High School Affiliated To Renmin University Of China (Ren Da Fu Zhong, (RDFZ))
qiujinjia@sina.com

Abstract

This paper makes an analysis of the status quo of the teaching of middle school mathematics in China and states the important role mathematical modelling plays in the reform of mathematical education in middle school as well as the urgency of the reform.

The paper also presents Ren Da Fu Zhong's explorations in mathematical modelling over the past years and the teaching models and accomplishments they have achieved in this field.

Finally the paper illustrates the author's viewpoint on the significance of the development of mathematical modelling in middle school and on its perspective.

1. Exploration

1.1 Background

The mathematical education in Chinese secondary schools is greatly in need of reform.

With the development of technology and science and the wide spread use of computers, humankind has come to the information age, in which technology of quantification and digitalization has been developed and applied broadly and quickly. "Mathematics and the science of mathematics have become the foundation for the development of high-level technology. The development of itself serves as a strong pillar for the boom of science and technology. Mathematics, showing from behind the curtain, is directly producing great profit for society in many respects." (Jiang, 1997). UNESCO designated year 2000 as World Mathematics Year, which shows that the world has acknowledged importance of mathematics. On one hand, mathematics is exerting its influence more and more greatly on all walks of life. On the other hand, the mathematical education of Chinese secondary schools has remained unchanged for several decades. Our traditional education of mathematics lays much emphasis on basic knowledge and skills training, but some problems are emerging such as teaching content detached from reality, ignorance of applications and lack of cultivation of ability of problem solving. In the process of reform, many experts and teachers are sharing the same idea that mathematical modelling is the solution to these problems.

1.2 Study: The change from participation of some of the teachers to all the teachers
       The change from general idea to a systematic grasp
The progress RDFZ has made in the popularization of mathematical modelling has laid a solid foundation for further.

       RDFZ has provided various opportunities for teachers' to grasp quickly the theory of mathematical modelling and to upgrade their ideas.

To undertake some research work
       RDFZ voluntarily undertook "the theory and research in the teaching of mathematical modelling in secondary schools" ---- a subsidiary subject of the key subject of China's 9th five-year Educational Science of national educational committee---" the theory and practice of mathematical modelling in secondary and primary education and the reform on mathematical education in 21st century". With the Principal in charge, over 20 teachers undertook different subjects and completed a series of teaching plans and papers concerning these subjects. Also RDFZ organized some symposia and seminars regularly, with a theme, a plan and keynote presenters. These activities have led to the teachers' improvement in their academic level and the advance of the research work in teaching.

To strengthen the exchange
       RDFZ is keen to exchange with other schools. Teachers are often organized by school to attend the open class in mathematical modelling of

other schools. Teachers benefited a lot from this participation and the following discussions right after the class. In addition teachers took part in some conferences and meetings in mathematical modelling.

To organize some activities

In 1997, entrusted by the Mathematical Association and the Contest Organization Committee, RDFZ hosted the 1st Mathematics Application Contest for Beijing High School Students and the following symposium. Some educators in mathematics like Mr. Jiang Boju, member of Academia Sinica of China Academy of Sciences, delivered the speech in the symposium, which was very beneficial and informative in mathematical modelling.

To consult the experts and professionals

During the past years RDFZ has invited some experts and professionals in mathematics modelling. They provided some helpful instructions to the research work of mathematical modelling in terms of theory, research method and presented solutions to the teachers' questions of how to develop the extra curriculum activities in mathematical modelling in secondary school as well.

1.3. Practice--- *efforts to find a way suitable to RDFZ*

At the early stage of the process, some thematic lectures or the Groups of Interest were organized for students. Some were open to all the volunteers; some were designated in a certain class. With the advantages of explicit principles, concentrative content and disadvantages of unsystematic content and poor participation, this organization could be used as the preliminary form of popularization as well as for some particular circumstances. It is suitable for Grade 9 who has a little knowledge of mathematics. Meanwhile they have a time limitation before their graduation. This form started in 1997 and lasted for two years.

We also tried the establishment of a required course for the pilot classes of Grade 10. Due to the inadequacy and short length of learning hours, this trial hasn't lasted very long, only one year.

Since 1999, we have adopted a new way that takes different forms for different demanders.

Some individual tutorials are given to the students in Grade 7 about their papers in application. Some papers have been published in professional journals. For example, one of them is 'About the Score

Scaling of Multiple Choice Questions' (*The Math Journal For Middle School Students,* Vol.12, 1999).

Some ideas about the applications of mathematical knowledge are presented to students in Grade 8. The mathematical application paper contests are regularly held during the summer and winter vacations.

The thematic lectures are held for the students in Grade 9 to obtain an introduction to the method and process of mathematical modelling. Some of the students gained remarkable achievements in the Mathematical Application Contest for Beijing High School Students.

Mathematical modelling is established as a selective course of two hours per week for the students in Grade 10 and 11. Generally three steps are adopted for this course--- the instructor's demonstration, the students' activities and the instructor and the students' discussion. The content is confined within the textbook. These selective courses result in an appraisal from the students.

The thematic lectures on the application of mathematics are also presented to the students in Grade 12.

The most important thing is that a proper involvement of applicable idea, applicable questions and applicable settings are embedded in regular lessons of mathematics through the junior and senior high school. This makes students aware that mathematical knowledge and the mathematical application are closely related organisms (RDFZ, 1997).

1.4 Experiences of the author

I myself am keen on the teaching of mathematical modelling in RDFZ and undertook the teaching of mathematical modelling in Beijing Mathematics Training School. In October of 1998, I presented a teaching demonstration of mathematical modelling in Haidian District. Over 140 teachers attended this demonstration, among whom were teachers from the Symposium of Beijing outstanding young teachers and some teachers from beyond Haidian District. This demonstration was my tentative exploration in mathematical modelling teaching. I would like to give a quick review of this demonstration.

Preteaching:

Based on the student's level and what they have learned, I chose *"research of the different price rules for the same kind of commodities"* as the subject of the research. I made an investigation in advance in several department stores and completed the entire process of the mathematical modelling. In so doing this research was proved to be feasible. Then I made an outline for the investigation for students. With this outline students were asked to make an investigation of the market. After that they were expected to build a mathematical model on the basis of their data and to present their data, the process of model building and their conclusion using an OHP.

While teaching:

Since the students were not very familiar with the process, procedure and method of mathematical modelling, I laid the emphasis of the lesson on the students' grasp of the general process of mathematical modelling and of the analysis of the calculated outcome. After the revision of the general process of mathematical modelling, some typical presentations of the students were made to illustrate their process and conclusions of mathematical modelling. Following that came the students' comments on each presentation about its merits and shortcomings. Finally the students would be asked to build a polished mathematical model according to the main procedures of the process so that the students could have a vivid and intuitive knowledge of solving a problem by applying mathematical ideas and methods. With the completion of the entire process of mathematical modelling, the students had an in-depth understanding of the mathematical modelling. To reinforce this result, the students were required to improve and polish their models and complete papers.

From a general viewpoint this class achieved the expected goal with a fruitful result. The students were upgrading their ability to analyze and solve problems applying mathematical ideas with great enthusiasm. They were so keen on their investigation and paper writing. I would like to quote a student's words from his remarks in his investigation report:

I think this activity so interesting and I am so happy to have finished an investigation and a paper on my own.

It is fun thinking that there are such a lot of interesting mathematical questions hidden in daily life. This mathematical assignment is the most attractive one among all during the National Holidays. When I was making investigation in different department stores I felt that I was performing something. I felt especially proud that those salesmen viewed me as an adult. After I was back home I was engaged in my first investigation report in front of my computer. I polished my report by adding some graphs and using different penmanship. This is my product,

my works, so I care very much about it.
  In one word, I have been totally involved in this job (Qiu, 2001)

These simple, plain words inspired me a lot. To render the students freedom to help them gain knowledge and improve their ability is the core of the education reform as well as all the teachers' wishes.

1.5.Accomplishments

We have achieved much more than the improvement of the scores and winning some golden medals.

Results of scientific research

In June of 1999, trusted by the department concerned, the subject committee and the audio-visual center of RDFZ made a film of the special subject *Secondary School Mathematical Modelling*. All the teachers of mathematics are engaged in the production. This film of special subject was designated by the Educational Committee as *"the Continuing Learning TV Thematic Lecture for the National Secondary and Primary School Teachers"* and was televised by CETV (China Educational TV Station). Over the past years of teaching practice and exploration, RDFZ has basically formed a systematic system and model of mathematical modelling and compiled *The teaching outline for the mathematical modeling course of the high school affiliated to Renmin university of china.*

Upgrading of teachers' quality

The practice of mathematical modelling has strengthened teachers' awareness of teaching reform and mathematics application, the idea of innovation and life-long learning. It also improved the teachers' ability in scientific research and strengthened their spirit of cooperation and teamwork.

With the guidance of experienced teachers, a group of young teachers of mathematics have grown up and actively engaged themselves in mathematical modelling activity by compiling books and papers and performing teaching practices.

The improvement of students' quality

The practice of mathematical modelling has fostered the students' creative spirit and the awareness of mathematical applications, cultivated their scientific attitude of respect of facts and linking theory with reality, and improved their ability of comprehensively applying what they have learned from books to problem solving. Through involvement in the whole process of mathematical modelling, they acquired a preliminary rational cognition toward raising and solving the questions, utilizing modern technology to search, collect and organize information and resources, conducting social investigation, and completing science-tech papers and oral defense. Meanwhile they learned how to cope with the relation between division of work and collaboration, individual and collective. All this will contribute positively to the cultivation of the students' non-intellectual quality and sound personality and the improvement of their comprehensive quality.

Some students have performed remarkably in all the Beijing High School Mathematics Application Contests. The numbers of awards places this school in the front. Some have been awarded First Places when in Grade 9. Moreover, some students have published papers on mathematical modelling in the professional journals such as *Mathematics Newsletter* and *Math Journal For Middle School Students*. One example of these is 'A Modelling On The Distribution Plan Of Motor Vehicle Lanes' (*The Math Journal For High School Students*, Vol.2, 2002).

Numerous students have chosen subjects, consulted materials and resources to complete plenty of papers on mathematical modelling in their spare time. *A symposium of mathematical modelling of the students of the high school affiliated to Renmin university of China* was completed in May of 2000, which contains some awarded papers in all the Beijing High School Mathematics Application Contests. With these efforts, we aimed at popularizing mathematical modelling among teachers and students and convincing them that mathematical modelling is not too hard to be accessible. Some questions do not demand advanced knowledge of mathematics to solve. Take a student of Grade 9, Feng Feng for example, he designed a new style envelope using his mathematical knowledge.

2. Viewpoint

Based on my previous experiences, this is a personal view to how to conduct mathematical modelling in secondary school.

2.1 Teaching Principal and Purposes

Teaching mathematical modelling is not merely a matter in college education. It should also receive attention in secondary education since this stage is the prime period for students' accumulation of knowledge as well as the crucial moment of the formation of students' world outlook and methodology. The cultivation of students' spirit of innovation and ability of practice weighs much in secondary education. Teaching mathematical modelling enjoys advantages to approach this purpose. So it is worth being popularized as an efficient way to the reform of traditional and conventional model of teaching and education. However, rather than aiming at learning new knowledge, mathematical modelling in secondary school needs to help students to acquire a systematic and thorough mastery of what they have learned. It needs to help them sense the interest, beauty and usability of mathematics by increasing their sense of applying mathematics through problem solving. In the activity of mathematical modelling, students can improve their ability in using tools, independent studying, team-working, social interaction and problem analyzing and solving as well as developing a rigorous, careful and persevering attitude toward science through data collection and materials consultation. They can also polish their view of innovation and ability of expression and writing by observing and analyzing the surroundings, proposing their own ideas and completing papers. All this cultivation needs to focus on beyond and after school. Students are supposed to benefit from it for all their life.

2.2. What is to be taught, How to teach it and How to launch the activity of mathematical modelling

What is to be taught?
        The subjects selected by the teachers should be characterized with being commonplace, easily comprehended, interesting and open. It will be better if the subjects are
---- available for students to investigate
---- available for students to resort to computers
---- with an outcome for practical use.
        The knowledge it demands should be
---- within what students have learned
---- related to other subjects such as physics and biology.
        Multidisciplinary questions can help students understand the relation between different subjects and are in accordance with the principle of college examination reform as well.
        Nevertheless, due to lack of resources teachers need to make a constant accumulation. I have put all my work over years into *the teaching*

outline for mathematical modelling of the high school affiliated to Renmin university of china.

How to launch the activity of mathematical modelling

----- to organize extracurricular activity centering on mathematical modelling, full involvement, using interesting and vivid applications makes this approach more suitable for junior school students.

----- to establish the course of mathematical modelling (such as selective courses and thematic lectures) to study the thought and method of mathematical modelling by selecting questions, analyzing and solving problems .

In the very beginning teachers can determine subjects and make an outline. Gradually it will be the students' turn to find subjects and solve problems themselves based on the teacher's trigger questions.

----- to infiltrate ideas of mathematical modelling and to adopt some applicable questions in mathematics class.

How to teach

A teaching model for mathematical modelling in secondary school can be established with three characteristics.

A. Subjectivism education

Subjectivism education is an educational way of developing students' initiative to learn by stimulating their sense of self-involvement.

In order to do so, we focus our teaching on students' characteristics. For instance, secondary school students tend to reveal their own ability and are subject to emotion. So we emphasized their individuality and provided them with opportunities to reveal their advantages. In addition, according to students' strong curiosity, broad interest and active thinking, we designed some open and active questions for students to tap their potentials. We usually infiltrate some ideas of mathematical modelling and questions into regular mathematics classes and let them realize that learning and applying mathematical modelling is not exclusive to college students. The preliminary mathematics can be applied to solve some questions around us. As to those who are unfamiliar with mathematical modelling, we helped them with their difficulties before shifting to mathematical modelling. In doing so we aroused students' interest in mathematical modelling and help them develop their subjectivism and initiatives.

B. Open teaching
Meaning to guide students to acquire mathematical modelling from daily life and practice to improve their ability of problem solving. For example, two students studied the print and design of textbooks.

C. Dimensional teaching
Meaning to take advantage of every means of teaching to produce a teaching multifaceted, multidirectional and multi-channel model.

2.3 Some problems for teachers to pay attention to
A. To upgrade teachers' ideas
Instead of drawing conclusions with teachers' demonstration and instruction, teachers can provide students with plenty of opportunities to tap their potentialities so that they can involve themselves in the process of problem solving. The teachers' role is to guide, inspire and evaluate.

B. To be aware of students' individuality and advantages and tap their potentials
Some students are not very interested in mathematics, but they are keen on military, sports and computers. Teachers should guide them from their interests and generally this will work well.

C. To pay attention to the progressiveness of the instruction
The subjects selected should derive from practice instead of ones made up by teachers. This consequently requires teachers of making analysis of feasibility.

3. Final comment
To sum up, although I have accumulated some experience and expanded my knowledge structure, I know I have a long way to go. My job is very ordinary, but facing the growing students I have no reason not to give 100% to my work.

References
Jiang Boju (1997) 'To make known the significance of mathematics to the general public world-wide' *Develop contest of mathematics application to improve mathematical education reform*, Committee of Mathematics Application Contest of Secondary Schools, Beijing, 7.

Qiu Jinjia (2001) 'On Mathematical Modelling Activities at Middle School Level' *Mathematics Newsletter*, 2.

RDFZ (2000) The High School Affiliated to Renmin University of China, *Summary Report on the Theory and Teaching Research of Mathematical Modelling at Middle School Level*, 11.

Section G

Panel Discussion

26

ICTMA 10 Conference Symposium

Chairperson: Peter Galbraith (Australia)
Panellists:
      Ken Houston (United Kingdom)
      Tomas Jensen (Denmark)
      Gabriele Kaiser (Germany)
      Yongji Tan (China)

Introduction: Peter

Welcome to the ICTMA symposium. We are of course a rather special group in that our membership is drawn from different academic and professional communities. There are some of us who are mainly mathematicians with an interest in education, there are some of us who are probably mainly education with of course an interest in mathematics. That's both a strength and a challenge, because it means that it brings together people with a variety of expertise and interests that one often does not get in conferences which, for example, might be all mathematicians or all education. I guess it also raises its own problems or issues, but what it really means is that because we come from disparate communities in that sense, and we meet only rarely, that we don't always get the opportunity to share and set the agenda for the future, so that the community as a whole can move forward. And this panel session is specifically to look forward and to address what we see as some of the emerging challenges for the community in the coming years.

We are going to organise things along these lines: There are three questions initially which I will just explain in a moment. What I am going to do is to give each panellist a maximum of three minutes to speak firstly to question one, then to question two, then to question three. I will then ask from the floor for any additional questions that you think are important to add to this initial list of three. I'll record these on an overhead and we will add those to the list that you see there. That will then form the basis for the discussion for the remainder of the time. You will be invited to respond, or add, or contribute to the dialogue on any of these original questions, in addition to what any of the panellists may have said, or to any of the new questions which might emerge at a later phase. At that point if you have questions you may direct them to the panel in general or a particular panellist and we will just allow the discussion to be as fruitful and as wide ranging as possible. The first question:

Given that the mission of ICTMA is to promote applications and modelling in all areas of mathematics education, what are the impediments to success?

It has been explained to me that not everyone may be familiar with the meaning of impediments. It means blockages – what gets in the way of what we would like to do. So if we can think of these, what are the blockages or the things that get in the way of what we might be trying to achieve. Secondly:

What are new challenges to the pedagogy (didactics), the teaching aspects of application and modelling? And thirdly

What are the most important research questions that we think the community needs to address in the immediate future?

So that's enough for me, at least for now, so I'll invite the panellists to speak – perhaps we could work across. Ken's sitting on the end, so I'll invite Ken to begin by addressing remarks that are significant for him regarding the first question on the overhead.

Ken:

Thank you Peter. Good morning everyone. I work in higher education and so I'm mainly concerned with the teaching of modelling, mathematical modelling with applications at university level. Now the impediments to success that I see in some universities – well I think there are two of them – and I'll look at both of these. The first is the conservative nature of many academic mathematicians. "The way in which I was taught is the way that I shall teach". For many people, this is their philosophy of teaching and this is what they do. I have observed this particularly in departments of pure mathematics. And they feel that the "definition, theorem, proof" method of teaching pure mathematics is the right thing to do. It worked for them and so it should continue to work for other students. But while this is an important and necessary part of higher education, of mathematics education, it is not sufficient. And we do, I believe, need to encourage such universities to embrace applications of mathematics and mathematical modelling to a greater extent.

I should say that over the eighteen years of the existence of ICTMA many of the universities in the United Kingdom have embraced the principles that we aspire to, to propagate the idea of modelling and applications in all our undergraduate courses.

Impediments are one thing, but finding ways to overcome them is another, and I believe I see the beginnings of a mechanism for overcoming some of these impediments. Recently, over the last two or three years within the United Kingdom, every mathematics department in England and in Northern Ireland (Scotland and Wales were considered at an earlier stage), underwent an assessment of the quality of their education. Out of that has come a bench marking statement. This is a document that specifies the benchmarks or the standards that should be adopted by all courses in mathematics, and within that is a requirement to include some applications of mathematics. The other impediment to success I see is the resource question. Many mathematics departments have reduced the number of faculty members, they have increased student numbers and with big, big classes it's very difficult to teach modelling in the way that we would like to teach it.

Yongji Tan:

Colleagues, I would like to say that each ICTMA I attend is a very successful conference. As Chinese participants, we have an opportunity to show you what's happened in China for modelling, teaching and applications. And we also learn a lot from foreign colleagues. What impressed me very much is that the foreign colleagues have done a lot of interesting work on evaluating, quantitatively or qualitatively, the modelling ability of the students. In fact that's also an evaluation of modelling teaching. In China in recent years, we have paid much attention to encouraging more colleges and a university to establish their modelling courses. Now more than five hundred universities and colleges have included modelling courses in their curriculum. We are at a stage to pay much attention to improve the pedagogy of modelling and application. I think the main problem here is the lack of qualified teachers of the mathematics body, so we plan to run some planning courses for teachers of mathematical modelling.

Finally, I would like to say some words about my research interest about the modelling teaching. I am interested in investigating the experiment, the role played by experiments and projects in modelling teaching.

Gabriele:

I think we are all aware that in some situations there is a consensus in the theoretical education debate that educational modelling should play a high role in math education. But if you look all over the world we find out that the situation is not satisfying and I think that we have to work on three different levels in order to overcome this situation.

I think first that we need more theoretical reflection. I would analyse the situation that there are not enough theoretical discussions going on – what we really want to achieve with applications and modelling in maths teaching either at school or at university level. We tend to emphasise the goals we have, which have been promulgated already ten or fifteen years ago, but we have not really thought a bit more deeply about what kind of goals we want to achieve. So I think in order to overcome this situation one approach would be to work theoretically.

The second level where I think we need more work in order to change the situation is the area of empirical research. I think we have to find out more about empirical obstacles, which prevent our students as well as our teachers from learning mathematical modelling, or teaching mathematical modelling and applications. I tried in my talk on Sunday [see Chapter 1] to describe empirical obstacles which might be in the minds of our students, and that was meant as research in order to overcome these difficulties. But I think much more work needs to be done in this area – the lecture of Sue [Lamon] we just heard in the morning is a part of this research as well.

And the third level I think we need to work on is the pedagogical level I would call it. It means we have to find out how we can convince our teachers to teach mathematical modelling and applications. The problem is not that we do not have enough material. I think that at least for Germany I can say that we have many examples published and a lot of material. The problem is that our teachers do not teach mathematical modelling and applications, and we have to find out what the obstacles are which prevent our teachers from teaching mathematical modelling and applications.

And so in a circle we come back to my first two levels. The need for theoretical reflection as well as empirical research in order to find out how to convince our teachers.

Tomas:

I also want to mention three impediments that are closely related to each other. Not surprisingly they are also closely related to what has already been said.

The first one is what I will call *syllabusitis* – which would be an attempt to say that there is an illness in the system to far over emphasis on syllabuses. This causes an immense amount of difficulties for the teaching of modelling, since modelling can be said to be an attempt to deepen the awareness and the understanding of our students instead of widening the number of concepts that we cover. So the *syllabusitis* problem is closely related to the second point I want to make.

That is, it is very difficult to teach mathematical modelling, simply because we don't have this secure thing as a list of concepts to follow. There are a lot of very difficult aspects in the teaching of modelling. I won't dwell on that because everyone here is aware of that, but that's not something that we can just research our way out of. We simply have to deal with the fact that it is, I believe, more difficult than teaching what we are more familiar with, pure mathematics.

And the third aspect I want to mention as an impediment is related to what you said Ken, about conservative mathematics teachers, and I want to maybe say that a little bit more provocatively than you did. That I think it's an impediment that almost all mathematics education, and research in mathematics education (in modelling?) is done by mathematicians. You said Peter that one of the strengths here could be that it brings together people who come from mathematics and have an interest in education and the other way around. But as I experience it, there is a strong emphasis on the mathematics part, and I will give two examples. The first one is that when we research in this community, when we research in the modelling process we tend to focus heavily on the parts of the process that are closest to the mathematics, the analysis and the mathematisation, and the purely mathematical results. Which means that we tend to lack emphasising the other parts of the process, and I think that is because they are further away from the mathematics, which is what most of us come from. The second example is from the teaching point of view. Many of you probably know of Mogens Niss [Roskilde University]

and some of the work he has been doing from a research point of view. I
tend to believe that the most important thing he has done for achieving the
goals of ICTMA is the set up of mathematics education at Roskilde, the
mathematical institution there. Simply because he has been successful in
setting up a mathematics teaching program where modelling has a very
strong place, and I tend to believe that that is a more important solution
than all the eminent research he has been doing.

Peter:
    We will come back of course to the issues that have been raised.
    The second question: **What are the (with the emphasis on NEW)
challenges to the pedagogy, the didactics, the teaching aspects of
applications and modelling?**

Ken:
    I'm not sure just how new these challenges are but they're certainly
timely and pertinent to me. One of the other hobby horses, or one of the
other main interests that I have is in the teaching of what is called key
skills. The skills of communication and working with other people and
learning how to learn and things like that. Now I believe that courses in
mathematical modelling are a very efficient and effective way of
developing key skills, because all the ingredients are in place. So I see that
as one of the things that we want to try and encourage across the
community. There is pressure on the universities in the United Kingdom,
and I'm sure in other parts of the world as well, to make sure that our
graduates do have these very important personal skills. The other thing
that I see as a challenge, but it's perhaps not new and has been around for
some time, is the integration of technology. We have heard some papers at
this conference of how people are doing this in schools and universities.
The practices are perhaps not wide spread and it is indeed quite a
challenge to see how we might integrate use of technology in our teaching
in an effective way, and I can see that as something that we need to look
at.

Yongji Tan:
    As I mentioned a few minutes ago, in China the students don't like
the teacher to teach modelling in formal classroom teaching. So, recently
we tried to use some experiments, mainly numerical experiments in the
classroom, and we also used some projects from real industry to improve

the teaching. So I think that's a new challenge--how to find a correct way to find projects and experiments. That's a new challenge for us I think.

Gabriele:

As I already mentioned in my first presentation I have made three points at another time. I think I would discriminate three different challenges that we have to face in the coming future. First I think, as already mentioned, psychological research is necessary on the teaching and learning of applications and modelling. I think we have to discuss whether something like 'part skills' are necessary in order to do modelling and the modelling process, or whether we need to emphasise meta-knowledge about the whole process of modelling. So I think we need much more psychological debate about knowledge needed for modelling and applications, in order to promote modelling and applications in our teaching. And I would like to refer to the very early work of Vern Treilibs in the beginning of the eighties about the formulation phase, and the modelling skills necessary in the formulation phase, which unfortunately has not continued. Now we think that in this area we need more research, and that would be one challenge.

The other big challenge I think is concerning the cultural dependency of examples and goals. We as mathematicians tend to accept or adopt the theory of Platonism which means that we consider mathematics as a subject which tends to be culture free, which means to be objective. There are of course many things. One theoretical feature of mathematics is to assume it to be culture free and objective. And I think much of the current research going on in the area of mathematics education has pointed out that this is not true, that there is this very strong cultural dependency of mathematics and of mathematics education as well. I think we have to consider much more the cultural dependency of mathematics and mathematics education in the area of applications and modelling. And I think we have especially to consider the needs of many multi-cultural students. We have, at least in Germany, and I think all over the world, many students who have a multi-cultural background. And we have to consider, especially in the area of applications and modelling the context, and where work examples come into play we have to consider the multi-cultural backgrounds of the students. Certainly the third level is something like a summary of these first two levels. I think they need much more theoretical reflection in the area of applications and modelling dependent on referring to the culture dependency of mathematics and mathematics education.

Tomas

The first thing I will mention is similar to what just started off with Gabriele. The formulation or problem posing part of the modelling process is not emphasised enough I think, and it is really a pedagogical challenge to find ways of dealing with that part of the modelling process, both for teachers and for students. And I think there is at least one major reason why we find it so difficult; mainly that in that part of the process it is absolutely essential to let the students be in control, and therefore we need to teach them to act autonomously, and that is a very new situation for mathematics teachers. Normally we are totally in control of what is going on, and we know far more than the students, but in this part of the process we have to give away this control, and it's a major challenge for us to find ways of dealing with that, and in teacher education and in teaching practice.

The second thing I find important to stress is the pedagogical challenge of finding ways of assessing modelling. If not for anything else then simply because everyone knows that if we don't change the assessment then we can't change the way teaching goes on effectively. And also here I can see one more reason why we find it difficult to find really good ways of assessing modelling, and that's because we have to change to a much less systematic way of assessing mathematics teaching than we have been used to. In normal mathematics everyone is used to simply giving marks for each different part of the problem, and then adding them up and adjusting a bit and that's a grading. When we want to assess modelling skills we have to act much more like people from humanities for instance, because we get reports handed in. We can't mark them simply by giving points. We have to see them as a whole and that gives it a much more subjective nature, which we are certainly not used to as much as they are in other areas of the school system.

And the third thing is technology but I don't want to dwell on that, that's obviously a challenge.

Peter:

From teaching to research - the third question: **What are the most important research questions?**

Ken:

This is of course a very personal answer. The people from the United Kingdom at this conference have presented some work on attempting to assess or investigate modelling skills using a rather simple test paper. One of the things I think we are finding is that some students have difficulty in discerning between what is good and what is not so good. And I think this is certainly a research question that I would like to follow up with my colleagues: "How do students learn discernment?" The other aspiration I have, whether this is research or philosophy I'm not sure, but I'm wondering can we find a unified theory for the learning of mathematical modelling. There is quite a lot of work in the literature about situated cognition, and about cognitive apprenticeship and things like that, and I think it might be useful for us to explore some of these, and to see if we can put together a coherent theory for the learning of mathematical modelling. These are the things that I would like to follow up on.

Yongji Tan:

I'm from Fudang University in Shanghai. After our hard work recently in our department, we have elementary modelling courses for sophomores, and we have advanced modelling courses mixing case study and experiments and projects, and also we have a seminar for the last year students, and they do some projects for a bachelor's thesis. So as I mentioned I am personally interested in investigating the role played by experiments and projects, and also some of my colleagues are interested to find some new fresh material to put into the modelling courses. Because the students in our university are really high level (because Fudang University is a very important university in China), they want to know more, and so that's a problem to find fresh material which would help the students to grasp the methods of the modelling process. And also this material can stimulate the student to love the industrial problems and do something like that. So some of my colleagues are interested in this field.

Gabriele:

This is a very personal question. I can only ask it personally. I think from my perspective empirical research very closely related to practice, educational practice or teaching practice, is very necessary. And I will just briefly touch on the different aspects that I will follow in the next few years and which hopefully will promote discussion.

First is something Ken you already mentioned - everything concerned with situated cognition. That means empirical research around the dependency of situation and cognition. How does the observation of the situation influence the cognition? The work I have presented here on context definitely belongs to this aspect.

The second area or the second aspect in which we definitely need empirical research is the question what we need to find out about how we can realise our goals in math education connected to applications and modelling in teaching practice. So we need practising teachers or student teachers who really carry out application and modelling based instruction. We have to find out how it works. So we need research closely connected to teaching practice.

And then the third direction already mentioned is the multicultural aspect. I think we have to consider empirical research which tries to find out how the cultural background of our students influences their thinking about mathematics, and their thinking about modelling and applications; especially examples and the influence of context when you have a strong cultural dependency. That's my deep belief.

And something which has interested me in the last ten years very much but I haven't connected very closely in the area of applications and modelling (but which I really would like to connect with a little bit more), is the area of international comparisons. I would like to know a little bit more about what's going on internationally in different countries I think. As I said in the beginning we tend to say there is international consensus, and then we say something which we all agree about, but which we have not really researched deeply! Where do we come from? Are there internationally accepted obstacles or maybe promotional aspects toward the teaching of mathematical modelling and applications?

Tomas:

Like you Gabriele I also want to stress my interest in the relation between research and practice in math education which is something where there is a far too wide gap I think, and I am very much interested in working with those kind of research questions. And my way of doing that is to try to direct my interests not to the 'why?' of modelling and applications, but to the 'why nots?', simply because that's the way it's often seen in the practice. They experience the problems of teaching modelling and applications and also when we watch practice we can see that it's not going on in very many places, so there is a need to research on the 'why not' question and we don't tend to do that very much. We tend to take the positive approach which has been researched for twenty years and there are many beautiful answers there but not so many to the 'why not'. Related to that is another interest of mine, namely a methodological one. More and more I tend to believe that we can use the mathematical modelling process also as guiding principles for our research. We could perhaps call it didactical modelling, simply to emphasise that we should end up having a connection to a didactical situation, not in the French sense of the word, but a situation that has didactical content-and not just leave it in the open air.

Then I have two more specific ones that I will deal with right now. One of them is, in what ways do the challenges (both for the student and the teacher and the researcher), vary throughout the modelling process? Perhaps that's a bit similar to what you mentioned Ken about working with this learning theory for modelling. I think we would need to work with different ones there, because I think as I mentioned before, the challenges in the outer part of the process are very different from the challenges in the inner part of the modelling process, both for the teachers and for the students. We need to be aware of that and not just think of it as one thing.

And the last thing I want to mention is also very specific, that is when we want to deal with these first parts, the formulation, the problem posing parts of the modelling process. When we do that what is the potential of starting to think in terms of competencies, more specific modelling competencies as a guideline? What are the difficulties for the teachers and the students and what can we hope to achieve if we start thinking in those ways instead of the syllabus way?

Peter

Thank you panellists for starting off the discussion so well. Before throwing the meeting open for general contributions to the discussion, you will notice we have here the three initial stimulus questions from which we have had our panellists responses. I am going to give an opportunity for people from the floor to add up to three more questions they consider essential, which may then widen the basis for further discussion. I would only ask one thing. Please if you do that, we would be only interested in questions that are distinctly different from any of the three that are there. In other words if your question can be addressed under one of those headings then the appropriate way to do it would be to ask it of the panel as we move into the discussion time. But I think it would be interesting if people in the audience feel that there are issues that are distinctly different from those that we have already used to stimulate the discussion, then let's consider that there's a vacancy for up to three such questions. Are there any such?

John Izard (Australia):
        How do we improve our communication with potential conference participants in order to increase participation?

Peter:
        Thank you John. Are there any other questions that we might add to that list? Up to another two. Going, going gone!
        I'm going to now invite you all to join in the discussion and I think the simplest way to do that would be if there is a sub-question or a point that you would like to add or seek further clarification from the panel, then please ask it and I think we'll just say ask it of the panel in general and the panellists among themselves can decide who will respond to that question. Otherwise you may wish to make an individual contribution by means of addition and of course please do so. So either further questions to the panel for further clarification or a statement with respect to one of the four questions that we have now before us. Who wants to start? Chris, thank you.

Chris Haines (UK):
        The panel helped us to understand barriers to modelling at all levels in mathematics and mathematics education. I wonder if they've got any views about how-what incentives there are for teachers and lecturers to change to include mathematical modelling in their teaching and in the curriculum. That is beyond simply the need to clearly articulate the goals of mathematical modelling, and making sure those goals are covered within the goals which are expressed in different ways for mathematics as a whole?

Ken:
        Perhaps if I could pick up on that to start with. We are talking about cultural change and institutional change. There are two types of incentives I think, there is the big stick and there is the carrot, and the big stick approach seems to be one that is being applied in the United Kingdom at the minute. As I mentioned in my opening remarks, we have

had this assessment of quality of teaching and there the investigators, 'the inspectors', (well it was a peer review process let me say), were concerned about clear statements, about the aims and objectives and clear linkages between the aims and objectives and the assessment processes. So there was pressure on academics to make sure that these linkages were there, first of all that the aims and objectives were clearly stated, and that the linkage with the assessment process was clear and well thought out. So certainly there was one mechanism that produced some change in the way in which academic mathematicians viewed their teaching. That didn't quite address the question of whether or not there should be applications in a mathematics course. Subsequent to this exercise the quality assessment agency of the United Kingdom have been establishing what they call bench marks for each of the subjects. Now the bench mark is a statement which tries to describe what's happening in the university sector, what's happening in courses and what the bench marking group believes is the minimum standard or the minimum curriculum to be followed for the average student, for the median student at honours level. And the mathematics statement is now available for consultation and the group has concluded the first phase of their work, which is to produce a document for consultation. This is available on the QAA website, I can give people the address if they want to look for it. But one of the things that it is recommending is that there should be some applications within every university mathematics student's education. It doesn't mention modelling but I think that's implicit in the idea of applications whether it is using statistical techniques to solve statistical models or situations or whether it's using differential equations or whatever to solve the dynamic types of models. So the big stick is one way of doing that and certainly there are moves in that direction, they are resisted, they are not popular and they require change, they require hard work. But certainly that is having an effect on the curriculum and ultimately on the way we teach and assess our students.

Tomas:

   I think in talking about mainly upper secondary level which is what I deal with most, I would say there are at least two incentives, two carrots so to speak. One is that for some teachers it can be an invitation to develop their teaching, and luckily there are quite a number of teachers who have started to want to change and that's one way of doing it. Some teachers find it very challenging, and it's simply a way to develop their own way of teaching, because they can easily see that they need to find new ways of approaching the teaching situation. The other thing for high school teachers, is that I think many have experienced that when they start doing modelling and applications, then they have a far better communication with the students, because it is an invitation to link the students' experiences with what's going on in the maths classroom. And once they start seeing that, then the atmosphere in the classroom can change simply due to this better communication among the participants.

Peter:
Thank you Tomas. Is there anyone among the body of the audience who would like to come in on this particular question?
In that case we will go on to the next question.

Delegate 1 from Audience:
Considering the fact that different faculties have got different requirements, do you think that a course, one single course of mathematical modelling is valid for all faculties, or different courses of mathematical modelling for different faculties? What do you think is best, one course of mathematical modelling for all the requirements of different faculties, or each faculty having a different mathematical modelling course?

Ken:
I would put forward the view that mathematical modelling is a way of life. It's a philosophy that people should adopt when they come to solve problems in the world. Now the sorts of problems that arise in the social sciences such as economics, sociology, the sorts of problems that arise in science, physics, chemistry and biology, or in engineering and technology are different. The problems are different; the mathematical methods used are different; the interests of the students are different; but nevertheless the process of mathematical modelling is to a large extent the same. I think, if resources can allow it, that it would be better for modelling to be embedded in each of the different subject discipline curricula. I think engineers would not be particularly interested in solving problems of economics and vice versa. So I think that while many of the ideas of the process of teaching, the process of doing the work, the way of life of the professional mathematicians (whether economists, engineers or physicists) are very similar, the problems they address and the methods they use are

different, and the mathematical methods they use are different. Therefore I would recommend distinct courses for the different subject areas, but with a common philosophy for the approach to the teaching.

Gabriele:

I would like to mention the results of the Third International Mathematical and Science Study in which it was found that the mathematical knowledge as well as the mathematical interests of students from a variety of backgrounds are very different. So I think it's just not suitable to address students who really have a big difference in mathematical background, (and that means mathematical knowledge), and in the interest and motivation they have to study mathematics, just through one unit course. I really think we have to develop courses that fit to the mathematical knowledge that the course requires for the problems you can tackle, as well as to the interests of the students. And I think the problems in modelling courses should of course relate strongly to the age and area of the study the students come from. So I think it just doesn't make any sense to have a course for engineering students together with science students or students from social science - I would strongly recommend the same as Ken just did.

Peter:

Does anyone else from the audience wish to address this issue before we move on?

Chris Haines:

Just briefly on that last point. Whilst within mathematics and modelling I would tend to agree with the panel responses, there are interesting experiments in England where for example journalists and civil engineers work together on projects. For example on the planning permission and the development of terminal 5 at Heathrow where there was substantial mathematics involved, but when you've got group work and different disciplines you can have members of a group contributing in different ways. So you can have a modelling experience in a varied group if you go across disciplines at certain levels. Now that's very hard to do within a school, I guess, but at a university level and particularly at the later levels in a university it's quite possible.

Ken:

Yes, just on that point Peter and Chris, the engineering faculty of my university have a common project in their final year and the project is a team project and they bring together a civil engineer, an electrical engineer, a mechanical engineer and a project management engineer to attempt to address a project which requires all of their expertise at that level. And that is at final year undergraduate level, and it seems to be very successful in getting people to contribute their expertise to a group to achieve a common goal.

Tomas:

I completely agree with you Chris, the experience we have at Roskilde with the mathematics study there is that some of the best projects that we do (which we do quite a lot), come from groups that have as their second subject next to mathematics, very different backgrounds. I've been in groups where I have economics as my second subject, and I was doing it with a biologist and a physicist, and that's a very interesting combination. But that's because we have a project at a mathematics department that we all agree that it is the mathematical modelling behind it all that we need to grasp. But my reason for being afraid of having all do the same course is exactly that there is then one less course, and the course tends to be very general in its nature where you can't really dig into anything, and then its just pure air.

John Izard:

Thanks Peter. There's a problem that is close to home where we should be applying our mathematical modelling expertise. Sue Lamon earlier on highlighted all the gaps in knowledge issues that have occurred during this conference. Perhaps we should be looking at how to improve our teaching of mathematical modelling by looking at the teaching behaviour. Perhaps it's easier for a student to tackle a task where there's a higher probability of success. Perhaps we should be looking at the success probability of our teaching tasks so that we can arrange our teaching so that students have some success experiences. I think that unless we address that issue then we're not likely to succeed in convincing teachers who don't teach by these methods to change their approach.

Peter:

Thank you John. Is there further audience comment on this question?

So I'll invite questions on a new theme.

Ken:

Would it be useful Peter to pick up the first question that John addressed and that is: "How do we improve our communication with potential conference participants in order to increase participation?"

This is a difficult one. I think it requires us, the enthusiasts, to enthuse our colleagues in our own neighbourhood, in our own country. There will of course be different routes and vehicles for doing this in each country. There are some in the United Kingdom that I can mention. For example the London Mathematical Society and the Institute of Mathematics and its Applications and various other professional and learned bodies publish journals and newsletters and of course it is important that the conferences be advertised in those, at least in the diary of the forthcoming events. It would also be useful I think, for us who have been here to try to write some sort of review of the conference, to say why it was useful to us, what we learnt, what it achieved and to try and publish this in our own journals in our own country, so that in that way we can try and increase awareness in the community at large, in our own country, of what's happening in this particular environment. I think too, we perhaps need to try to convince people that are not yet convinced that the teaching of mathematical modelling is important and useful. I would like to see this included in the pre-service or the early service training of new university lecturers. The standard method for entry to a university lecturing job is to complete a PhD,. and perhaps a post-doc. for a year or two. Now in many instances these students will have been doing modelling in their work, but when it comes to teaching they're not perhaps ready, or well enough informed to develop the teaching methodology. So, when we come to the early in-service teacher training of our university teachers, then this is something that should be included in that. So another method of disseminating information about this community, about ICTMA, is through some of these vehicles. There is another vehicle in the United Kingdom in the Undergraduate Maths Teaching Conference which is an annual conference and if people who go to ICTMA can go there and tell people about it, that's spreading the word as it were by word of mouth. We can of course try and advertise through our web page and things like that.

But personal contact and personal encouragement seem to me to be the best ways of increasing participation. Another way, which Gabriele I am sure will mention, is the idea of trying to find funds for young PhD's or young post-docs in order to attend conferences like this.

Gabriele:
        Of course that's an invitation to emphasise a little bit more. It is of course a deep concern, and I have already emphasised this at the last few conferences, but it hasn't succeeded to bring in enough of our PhD. students. So first of all we need more PhD. Students, at least in Germany, and I'm not sure what's going on in China or in other states but I think at least in Germany we need more PhD. students, who do research about applications and modelling, and the implications either for university level or for school level. And we have to bring in these PhD. students to our conference. And we discussed at the meetings of the Executive Committee several things about how we could promote our PhD. students, and I am very deeply convinced that Sue will do a very good job at the next conference to facilitate our young researchers to come to the conference. And then I think there are two other aspects I would like to mention. If we consider more the multi-cultural background of applications and modelling, and the cultural dependency as well as the international aspect, I think we can only do that if we improve the communication to participants who are in the country from which we are. So if we consider these two aspects I have mentioned, more than we have done so far, it does leave room for thought for the task John has given us.

Peter:
        Yongji Tan I wondered if you would like to share your perceptions of the situation in China with respect to continuing to keep ICTMA rolling along?

Yongji Tan:
        Yes. Just as I said that in China also the problem is the lack of qualified modelling teachers. Recently all the universities in China enlarged their enrolling student numbers, so the lack of teachers is a serious problem, especially for the modelling teacher. You see, usually the students, the PhD. students who learn pure maths cannot very easily teach mathematical modelling, so we have to give them some special training.

So we are paying much attention to running some summer courses or something like that for the teachers of mathematics modelling.

Peter:

Does anyone from the audience wish now to add further thoughts to the question on view? It's a very important question indeed.

Bahadra Man Tuladhar (Nepal):

I think I am the only participant from an underdeveloped or developing country in this conference, so if in the future the venue of the conference becomes something like Nepal, a developing country, in ten years there will be the possibility of bigger participation from neighbouring countries like India, Pakistan, Sri Lanka. There are a lot of people, I think, who would be interested in this type of mathematical modelling and applications, and we see that for the past few years all the conferences have been held in bigger countries, bigger cities and they have a group of people who have been participating mostly in this conference. So if you want to have a big audience you should come to Nepal. So it is my offer to offer that a conference can be held in Nepal for all the persons who are here. Thank you very much.

Peter:

Further comments?

[There then followed several suggestions concerning ways of exemplifying practice, and of sharing ideas about modelling and its teaching. A summary of this discussion follows.]

(Delegate 2 from Audience and Gabriele)

It would be very helpful to observe classroom teaching of modelling. This could be assisted through the production of short videos (of good and bad practice), which could then form the focus for discussion and analysis. It may be desirable to provide corresponding transcripts - in English.

(Marta and Ken)

At future conferences it would be useful to have different groups working on different issues such as these and bringing conclusions to a later meeting. This general emphasis could be helped by shifting the balance of conference presentations towards including more workshops, and group discussions on a common theme. These ideas have been suggested in the past but not implemented for a variety of reasons, including the necessity for delegates to present formal papers in order to receive funding to attend.

(Gabriele, Tomas and Peter)

The Kassel and Roskilde conferences tried versions of workshops but numbers remained small, and at Brisbane workshops for secondary teachers were included and repeated. It might be useful to run parallel workshops (90 mins), say one on teaching practice and one on examples. As not all delegates may favour workshops it might be useful to run workshop sessions in parallel with other presentations, rather than in parallel with each other.

(Sue and Gabriele)

For ICTMA 11 a day is planned to a site (industrial, bio-medical, secondary and elementary teaching) to enable participants to engage with the setting and write joint papers reflecting on the experience, and on how things are done in their home countries. There would be additional benefit if discussions with people working at the sites could address their views on the corresponding mathematical education desirable for students. This occurred at ICTMA 3 in relation to a visit to a Volkswagen factory.

Peter:

Right, we are now able to take just a few more questions, so is there a separate issue that some of you would like to raise, or we can always revisit an issue that you feel is incomplete from your point of view.

Delegate 2 from China:

I think I want to pose another question. In recent years, as we know, the Internet has developed very fast, and sometimes we are told that maybe in the future a lot of the teaching and the education will be on the Internet. Maybe some day all of the courses can be found on the Internet, and also maybe if this is the case it will be worth a student going to the website for their courses. So that maybe some day, only a few teachers will design the website- and I just got unemployed! Of course I don't hope that's the future, but I want to know the comments and ideas from the other people. O.K.

Peter:

Thank you. The opportunities and challenges being presented by the Internet of course are going to expand, only expand. I wonder whether the panel would like to respond to either the specific question, or their own perceptions of how the Internet and the Web might impact on ICTMA.

Ken:

One of the requirements of my faculty at the University of Ulster is that we put all of the information, or a good lot of the information of each of our courses, of each of our modules, on the Web. So that our students can access these from the computing laboratory or from home or wherever they can have an Internet connection. So each module, each course would have its aims and objectives, its content, its assessment methods and other things as well; such as the modelling exercises in my case, the modelling exercises that I set my students, and things like that of interest. The assessment criteria that we use in assessing project work and presentations and those sorts of things. There are some photographs for example too of students at a poster session, when they were presenting posters of their work. So there are some examples scattered about in the world of information that might be useful to people who want to go looking for it. What we need I think is some sort of index or some sort of data base, that would make it easier for us to put links to all these pages on the one page, and that might be an interesting job for someone at this conference to try and take up.

Peter:

There is time probably for one or two more issues to be raised or others revisited in new ways. Thanks Werner.

Werner Blum (Germany):

I would like to suggest two programs for the ICTMA community. Programs are both about development and research, but one is more development oriented and one is more research oriented. The first, more development oriented, is the subject of teacher education. We have identified as one of the major impediments the lack of knowledge, of experience, and the narrow beliefs of teachers. So we all know that teacher education is one of the most important challenges, and this is an enormously huge task that some people might not have thought of because we have millions of mathematics teachers in the world. Millions, not only hundreds of thousands, and I could imagine a permanent forum where we exchange ideas on how to really reach not only a few but a big amount of teachers to change their teaching in the ways we would like them to teach. So a permanent forum for exchanging ideas and conceptual teacher education is the first suggestion.

For the second I would underline some aspect that has already been discussed. That is that one of the main findings of research in education has been the context and situation specificity of learning and using mathematics. And that applies to modelling and applications and everything that we discussed here especially. I would like to see more research into various ways and modes of conceiving the teaching and learning of mathematics to cope with this inherent problem. That means how to organise and conceive the learning and teaching of mathematics, so that students become able to translate between mathematics and the real world context. I would imagine a permanent forum where we can exchange ideas especially for this aspect in the core, in the heart of what we are doing-the context specificity of learning, using ,and how to cope with that problem.

Peter:
Response to Werner's suggestion from the panel or audience?

John Izard:
I think that there's a lot to be said for what Werner said. I support it wholeheartedly, but I make one research recommendation. That is when we are conducting a study into whether something works or not we need to know at least two points--it can be before and after, or it can be before, during and after, but we have to demonstrate change and that has implications for assessment. And you can't measure the before with one ruler and the after with a totally different ruler. They have to be the same ruler. So it would seem that to assist what Werner said we also need to consider better ways of describing and reporting progress, and better ways of reporting the research that we do, so that others can build on it.

Gabriele:
I would like to comment on the last two contributions. First to you John. I think what we need in order to avoid the measuring with two different rulers is discussion about the methodological standards of empirical research. We need not only to carry out and present empirical

research in the next coming ICTMA Conferences, we need to discuss the standards of empirical research in order to avoid the usage of different rulers for the same thing. That's something that we have to consider in the future. And secondly I would like to comment on Werner's recommendation to have a discussion about teacher education. I think it of course one of the main problems we are facing in the coming future: How can we change teacher education? And it's not a contradiction it's just an addition. We have to keep in mind when we discuss about teacher education in different countries that we have to consider that teaching practice is very different in different countries. We teach in a certain way the basics of mathematics, that's of course similar all over the world, but even the notion of derivative or the rule of proof, are so different in the different countries. There are big variations and especially concerning the notion of applications and modelling, and the conception about modelling questions is very different in the different parts of the world. So in a certain way, before we can have a discussion about how we can change teacher education internationally, we need a good understanding of what's going on, what's our comprehension of mathematical modelling and applications in the different parts of the world. How mathematical modelling and applications is taught in different parts of the world, and I think we just have to do the first step before we can do the second step.

Peter:

Is there one final comment or question. We have just time for one.

Delegate 3 from Audience:

I would like to underline what our Chinese colleague told us half an hour ago. What we need are concrete examples well worked out which are stimulating, and which can be adapted in the different cultures. I do not completely agree with you Mrs. Kaiser if you underline this distinction too strongly. I agree in a certain amount, but there are a lot of problems that are common. We are a small world. Pollution, traffic just to mention a few key words. I think this is a problem for London as well as Beijing as well as Zurich and many other places in the world. So what we are still lacking is concrete, nice problems with experience behind them perhaps in the ideal case, so we do not travel in the dark at the very end; that the students have success as John told us. I think that is one of the basics for me, and just talking too much on a meta-level about philosophies (well it's a philosophy, I agree with Ken) sure, that's fine, but we need concrete problems, and it is hard to get them. You can have kilos of calculus with some nice examples and a lot of rubbish. Well why don't we cultivate that? I would be happy if we could just reach that goal.

Peter:

I think we give the panel one last opportunity to make whatever points they wish.

Tomas:

I was about to say that I am a little bit worried about all the emphasis on finding good problems in the group of teachers, because that is part of the problem of not letting the students work with the entire process. I would rather emphasise that we should work on teaching the teachers how to let the students do some of the problems. And in that sense it will be very local because it will be taken from the students' experiences. So I am a little concerned with the idea of saying that all we need are good problems, because then we will miss a very, very important part of the students competence.

Peter:

Thank you Tomas. It is probably good that we finish on a point where we have differing opinions because that keeps everything going and all of us on our toes.

I think all good things come to an end and I would like to firstly ask you to join with me in thanking the panellists for getting a discussion going that has raised so many issues that I think we can all take away. And the important point of course is not just that we take them away with us but that we work at them up here, and we work with each other and that we do indeed move everything forward, as was the intention. I would also like to thank you particularly, the audience, because I think your contributions have been magnificent also, and this has been for me a very stimulating hour and a half. Would you now join with me, please, in thanking the panel.

Notes